SHOPPING, PLACE AND IDENTITY

Why d ~~~~~~~~ upon their benefits as safe environ-
ments ~~~~~~ ~~ shoppers show such an aversion to shopping
with t ~~~~~~~~~~s? Why do some ethnic groups identify strongly with particular shopping centres but not others? Why do department stores come to
characterise certain shopping centres and how does this help people define
themselves in terms of class?

These, and similar questions, are explored in *Shopping, Place and Identity*.
Presenting a unique study of shopping, the life of shopping centres and the
nature of shoppers, this book offers new approaches to understanding the significance of place and the construction of identity. From an historical and thematic
survey of the nature of consumer societies and their implications for identity,
the authors examine the commercial and historical background of two London
shopping centres – Brent Cross and Wood Green. Drawing on primary research
which combined ethnographic work on shoppers from particular streets, with
the use of focus groups and survey questionnaires, the authors examine particular issues that arise in the action of locating identity and class through shopping.

Shopping, Place and Identity presents a firmly grounded study that will comple-
ment the more speculative writing about shopping, place and identity that has
developed in recent years. Engaging with key debates in contemporary consumption and identity studies, the authors show how the materiality of place operates
within the dynamics of identity in the modern world.

SHOPPING, PLACE AND IDENTITY

Daniel Miller, Peter Jackson,
Nigel Thrift, Beverley Holbrook
and Michael Rowlands

London and New York

First published 1998
by Routledge
11 New Fetter Lane, London EC4P 4EE

Simultaneously published in the USA and Canada
by Routledge
29 West 35th Street, New York, NY 10001

© 1998 Daniel Miller, Peter Jackson, Nigel Thrift,
Beverley Holbrook and Michael Rowlands

Typeset in Galliard by
Florencetype Limited, Stoodleigh, Devon
Printed and bound in Great Britain by
TJ International Ltd, Padstow, Cornwall

British Library Cataloguing in Publication Data
A catalogue record for this book is available
from the British Library

Library of Congress Cataloging in Publication Data
Shopping, place and identity / Daniel Miller . . . [et al.].
p. cm.
1. Shopping centres–England–London.
2. Shopping centres–Location–England–London.
3. Consumers–England–London–Attitudes.
4. Social classes–England–London.
5. Brent Cross (Shopping Centre: London, England)
6. Wood Green (Shopping Centre: London, England)
I. Miller, Daniel.
HF5430.6.G7S48 1998
381'.1'09421–DC21 97–35292
CIP

ISBN 0–415–15460–X (hbk)
ISBN 0–415–15461–8 (pbk)

CONTENTS

ILLUSTRATIONS

Plates

Figures

TABLES

PREFACE

In 1997, Brent Cross Shopping Centre celebrated its twenty-first birthday; it had officially come of age. Described in the *Independent* newspaper as 'the mother of malls' (9 March 1997), Brent Cross was Britain's first purpose-built regional shopping centre. Since its opening in 1976, Brent Cross has been overtaken by several bigger and more spectacular malls (Merry Hill, Meadowhall and Lakeside among them), but its claims to have been in some sense 'the first' have rarely been challenged. As such, it has attracted both praise and blame. Accused of destroying the traditional English high street, wrecking the environment and replacing freely accessible public places with sanitised and tightly patrolled private space, shopping centres such as Brent Cross (and their latter-day successors) have none the less proved extremely popular and financially successful places. At £300 a square foot, rents in Brent Cross are as high as in London's West End, offering consumers a safe and climate-controlled alternative to the perceived dangers and unpredictability of city-centre shopping. Over the years, Brent Cross has become an accepted part of many consumers' weekly (and in some cases daily) routine, no longer seen as a spectacular symbol of modernity and progress. Brent Cross has responded to the competition from more recent developments by undertaking a multi-million pound facelift, letting in more daylight, increasing the amount of free parking space and generally sprucing up its appearance, aiming to attract a younger generation of shoppers as well as those who grew up with the centre.

As Britain's first, Brent Cross is now its most fully established mall and those most concerned with the latest retail developments have moved their focus to other sites. As a suburban shopping centre, Wood Green Shopping City was also seen as innovative in its day. It was launched with high hopes as a combination of local political ideals and commercial plans. It was intended to ensure that the area would be able to maintain its claim to being one of the major suburban retail centres in North London. By the time of our fieldwork (in 1995) these sites had clearly lost any such sense of excitement and appeared merely as two of a mass of such suburban shopping centres. Becoming 'ordinary' meant they could serve as the foundation for our project. What we were seeking to establish through this research was less the 'meaning of the mall' and rather more the diverse and often contested meanings of these shopping places and their relationship to the identity of those who shop there. Rejecting the ungrounded theorising that has tended to dominate recent work on consumption (for reviews, see Miller 1995), we focus on these two particular places, contrasting the experience of shoppers at Brent Cross mall with the less prestigious 'Shopping City' at Wood Green. Much of the literature on shopping malls seems to start with issues posed by theorists such as Baudrillard, Benjamin and Bauman, transferred to the most spectacular and vanguard malls. Although we have been influenced by these same debates we have concentrated on shopping centres which have had time to become well-established features of the suburban landscape and our main source has been the people who shop there. Our theoretical concerns also tend to be less 'state of the art' and more attuned to questions of social identity and context that matter demonstrably to those concerned.

Our approach has combined traditional questionnaire surveys at the point of sale with focus group discussions in the surrounding neighbourhoods and a year-long ethnographic study of the area around a single street. Combining methods and approaches from geography and anthropology has, we hope, enabled us to avoid the academic condescension of much recent research, grounding our understanding of contemporary consumption in the lives of 'ordinary consumers' and letting their voices be heard through transcriptions from our focus groups and ethnographic observations. Such methods are appropriate to our understanding of the nature of contemporary consumption which, as we have argued elsewhere (Miller 1987; Jackson 1993), is a social process that goes well beyond the isolated act of purchase into cycles of use and re-use as the meaning of goods is transformed through their incorporation into people's daily lives. In what follows, we outline and attempt to justify our argument that shopping does not merely reproduce identities that are forged elsewhere but provides an active and independent component of identity-construction. We document the history of the two centres, contrasting the commercial imperatives that led to the development of Brent Cross and the wider political and community history that lay behind the development of Wood Green Shopping City. We outline our chosen methodology, highlighting the advantages of different methods for securing different kinds of data, exposing the frequent contradictions between what people say about shopping (often citing altruistic concerns for marginalised social groups or for the environment) and what they actually do (which is more often motivated by prosaic issues such as convenience or household interests).

Although the project was originally entitled 'consumption and identity' it will be evident in what follows that the results are rather different from the kind of work that has tended to follow the use of the term 'identity'. Very little of our material relates to the struggles of individuals with a sense of who they are, isolated from wider social and cultural contexts. At first glance the topics of our chapters may appear conventional – nature, ethnicity, class, family and so forth. But in reading the details it will be found that none of these labels ends up looking much like their colloquial use. If anything, it is contradiction which emerges as the starting point of most of the discussions: people who seem to hate shopping as families yet are drawn to places devoted to family shopping; the elderly who affirm one opinion in public and another in private; a sense of nature that is the result of carefully contrived artifice. It is in trying to account for these contradictions that we believe our material provides original and unexpected insights into the use of shopping centres. As a result the sense of identity is also expanded to incorporate the subtleties of its dependency upon context.

Similarly this is a study that tries to take advantage of the different perspectives of those who took part. Originally formulated by an anthropologist (Daniel Miller at University College London), the project became an inter-disciplinary one through the involvement of Peter Jackson (Geography, University of Sheffield), Nigel Thrift (Geography, University of Bristol) and Michael Rowlands (Anthropology, University College London), plus our research fellow, Beverley Holbrook (now at the School of Social and Administrative Studies, University of Wales, Cardiff), who had previous experience of using both quantitative and qualitative methods. As a result, the shopping centres have been analysed here in terms of locality and the use of space and place, but these terms are given substance by the variety of data sources and methods used, ranging from the results of questionnaire surveys to observations based on participation in shopping. The narrow range of questions that can be used in a relatively short questionnaire is complemented by the long period of acquaintance that forms an ethnographic relationship. Equally the small group of informants that comprise an ethnographic study can be generalised with more confidence thanks to the focus group and questionnaire evidence that pertains to a much wider spectrum of people.

This approach to the everyday nature of contemporary consumption is consistent with one of our key findings: that shopping is an investment in social relationships, often within a relatively narrowly defined household or domestic context, as much as it is an economic activity devoted to the acquisition of particular commodities. While our respondents rarely shopped as whole families (and those who did so rarely enjoyed the experience), we found social relations within the family to be the dominant context of contemporary consumption. The emphasis on family also underpinned our findings about people's understanding of neighbourhood change and the all-too-common racialisation of people's fear of change. We show that the discourse of contemporary consumption is often expressed in highly racialised terms, from the scale of the neighbourhood to the nation, sometimes mediated through notions of dirt and pollution. The carefully controlled environment of shopping centres and malls provides consumers with a haven from the perceived dangers of high street shopping and the risk of unplanned encounters with various (often racialised) Others. As such, the privatisation of space within shopping centres and malls provides a solution to the now widespread fear of public space, with closed-circuit television and other visible means of improved personal security adding to the sense of risk-free shopping (at whatever cost in terms of social and spatial exclusion).

Consumers' fears about the increasingly 'artificial' nature of contemporary shopping represents the other side of their desire for a carefully managed and crime-free shopping environment. Again, we argue that these feelings are as much about the social context of shopping as they are about the physical setting of the shopping centre or mall. When people yearn for a return to 'personal service' or support current trends for opening up enclosed shopping centres to 'natural light', we suggest that they are at least as concerned about the increasing artificiality of their social relationships (and in particular the perceived materialism of their children) as they are about the physical environment itself. Indeed, our focus group and ethnographic evidence found very little support for any form of 'green consumerism'.

We also suggest some new ways of theorising the link between ethnicity and consumption, mediated through the differential use of public space. Groups such as West Africans, who retain their links with their country of origin, or Afro-Caribbeans, whose public identity is circumscribed by the 'host community', tended not to use public space as an active part of their identity construction. Others, whose identities are much less 'fixed', such as Jewish mothers or Cypriot youths, made much more conspicuous use of public space in, respectively, Brent Cross and Wood Green. In this sense, the choice to shop in Brent Cross or in Wood Green, or to visit John Lewis or one of the discount 'cheapjack' stores, was as much about the construction of a particular (classed, raced or gendered) identity as it was about the supposed economic imperatives of price or quality.

The study was funded by the Economic and Social Research Council (award number R00023443) and ran from October 1993 to March 1996. We would like to acknowledge the assistance of the shopping centre managers at Brent Cross (Mr J. Bremner) and Wood Green (Mr H. Rudd) for providing access to the centres and for consenting to being interviewed. Alison Clarke undertook parallel research in North London in the same area as Daniel Miller's ethnography, focusing on other aspects of household provisioning, broadening the significance of the findings reported here. We have also made use of focus group discussions carried out by Vicky Knight under the supervision of Michael Rowlands. We should like to thank Zoe Phillipedes and Geoff Southall for help with data collection and Kit Kelly for administrative assistance. Presentations from the project were made at over 40 seminars and conferences and we would like to thank all those who participated in the discussions which helped shape our analysis. Finally, we would like to thank all those informants who responded to the questionnaires, formed part of our focus groups and who gave their time to the ethnographic enquiry. All of them must remain anonymous, but it is their contribution that made it possible for this research to take place.

1

CONSUMPTION AND SHOPPING

Introduction

Not so long ago, consumption was an academic outcast, rarely mentioned except in passing by any but a few authors who had usually stumbled across the work of writers such as Simmel and Veblen. Then came a period of expansion which, not entirely accidentally, coincided with a major consumer boom in many countries around the world. This period of expansion produced a number of canonical studies – works such as Douglas and Isherwood's *The World of Goods* (1979), McKendrick, Brewer and Plumb's *The Birth of a Consumer Society* (1982), Appadurai's *The Social Life of Things* (1986), Miller's *Mass Consumption and Material Culture* (1987), and Campbell's *The Romantic Ethic and the Spirit of Modern Consumerism* (1987) – which became the intellectual basis of the study of consumption. Diverse as these works were in character and style, they all agreed on three things: first, the paucity of theoretical or empirical research on consumption; second, the diversity of the social relations involved in consumption which made the category into, at best, a catch-all and, at worst, a confusion; and third, the need to consider consumption through many different kinds of social relations: gender, kinship, ethnicity, age, locality, and so on.

We are now faced with the fruits of a first generation of empirical studies of consumption, all the way from collecting to car boot sales and from catalogue shopping to party selling (Clarke 1997; Crewe and Gregson 1997), and with a profusion of theoretical frameworks, all the way from psychoanalysis to pragmatism (Bocock 1993). The problem is no longer that consumption is an unknown topic but that it is, in some senses, known too well: the unorthodox has become a new orthodoxy with all the problems that entails.

Of course, this trajectory is hardly unique. A number of other recent academic subcultures have followed much the same path, for example media studies and the sociology of scientific knowledge. Academic subcultures like these can even be characterised in some of the same ways as the study of consumption. They are fundamentally interdisciplinary. They are unsure of their exact focus; therefore they debate endlessly their central terms. And they have come to be

1

seen as particularly concerned with different kinds of knowledge and with the nature of the object.

How, then, can we understand modern consumption studies, and, most especially, the place of shopping as a crucial element of such studies? This chapter is a critical review of work in this field. To this end, it is in four sections. The first is a brief history of the study of consumption in three stages, highlighting the issues raised by each stage of work. The second then considers shopping itself. Here, the concern is both with the sheer diversity of approaches to shopping that are possible and with beginning to develop the framework which informs the work in this book. The third section then considers the issues of place and identity as vital determinants of modern consumption. In the final section, the four different threads of consumption, shopping, place and identity are brought together again through a consideration of the literature on shopping malls.

A brief history of the study of consumption

Writing a brief history of modern work on consumption is not easy. Studies of consumption have taken diverse approaches to an almost bewildering set of topics, all set within a number of disciplinary frames, each with their own procedures and protocols (Miller 1995). This section identifies three main stages of work, a first stage which covers the period from the 1960s to the late 1970s; a second stage which covers the period from the late 1970s through to the early 1990s; and a third stage which covers the most recent period of time. As with all such chronicles, this one is necessarily rough and ready.

The first stage

Let us begin by harking back to the early days of research on consumption. Why did 'consumption' become a rallying call for so many researchers? Four main reasons come to mind. First, a whole new landscape of consumption was coming into view: not just the supermarket (in Britain chiefly a product of the 1960s) but the retail warehouse park and the shopping mall (of which Brent Cross was the first real British example). Second, there was the implicit opposition to production. Consumption could therefore stand for many things – as an implicit critique of what were perceived to be productionist approaches (such as in Marxism); as the mark of a shift in the nature of production towards new times; or as an index of the decline of production-based working-class cultures and the rise of consumption-based middle-class cultures. Third, and relatedly, consumption was a way of gently introducing concerns about culture into social sciences still often dominated by approaches based on political economy. After all, consumption was still recognisably 'economic', wasn't it? Fourth, consumption, and especially advertising, provided a playground of interpretation for intellectuals in the humanities who, through the medium of cultural studies, were moving into the social sciences. These concerns of the humanities came together most obviously in the work on subcultures which were often identified with and through particular consumer objects (Willis 1975; Hebdige 1979).

The second stage

The second stage was one in which the study of consumption came to take on its own independent dynamic, becoming a recognisable subfield of a number of disciplines. Independence was declared in a number of ways.

First, consumption was cast adrift from production. Consumption became a world of its own, bolstered especially by the work of Bourdieu and de Certeau who became the all-purpose patron saints of consumption, with Bourdieu's (1984) consumer categorisations ameliorated by de Certeau's (1986) emphasis on amorphous, dynamic and flexible consumer 'tactics'.

Second, consumption became deeply implicated in discussions of the construction of subjectivity, most especially the construction of self and identity (Willis 1991; Nava 1992). Using a variety of theoretical frameworks, from social constructionism to psychoanalysis, consumer objects could be positioned as key elements of the construction of a whole range of selves and identities and most especially sexual and ethnic identities (Wilson 1992; Reekie 1993; Jackson 1994; Swanson 1995).

Third, practices which could legitimately be regarded as aspects of consumption proliferated. Consumption came to include practices like consumer festivals (Miller 1993), collecting (Belk 1995), catalogue shopping, new age shopping, and the like.

Fourth, an accepted natural history of consumption took shape which, identifying consumption as a key characteristic of modernity, described an arc from the arcades and department stores of Paris through to the shopping malls of the United States. Most specifically, consumption was interpreted as a part of the specular moment of modernity (Bowlby 1993; Pred 1995). Vision becomes the key sense because western societies are characterised by an excess of display which has the effect of concealing the truth of the society that produces it, providing the consumer with an endless supply of images that can be understood as either detached from the real world of real things – as Debord (1966) implies – or as simply working to efface any trace of the symbolic, condemning the consumer to a world in which everything can be seen but nothing can be understood (Cooke and Wollen 1995). The connections to the masculine gaze were quickly made (Bowlby 1985).

Then, fifth and finally, the study of consumption becomes increasingly integrated with and tied to spaces and places. In line with a general increase in interest in space and spatial metaphors, more and more attention was paid to particular consumption spaces which were no longer seen as just passive backdrops but as spaces with their own properties which could intervene in the

3

construction of difference. These spaces could therefore be studied for their own sake and not just as examples of more general processes (Glennie and Thrift 1992).

These five articles of a newly independent constitution formed the core of research into consumption in the second stage. However, like all such articles, they have proved open to revision. These revisions have formed the substance of the third stage. In this third stage, what we see is a growth in work on consumption, a parallel extension of work on consumption into new areas and, not surprisingly, the growth of doubts about previous writings on consumption as the results of this new research are digested.

The third stage

As in the second stage, there are five main issues. The first is, perhaps surprisingly, production and distribution. Many commentators feel that second-stage writings too often ignored the role of production and distribution, leading to both a lack of emphasis on the role of producers and distributors and an equal lack of emphasis on the role of consumers' choices on producers and distributors. Two particular illustrations of these lacunae are in order. First, there is the role of the workforce in shops and stores – from sales assistants to security staff to the managers and administrators, which is widely ignored even though they are an integral part of many forms of consumption. For example, du Gay (1996) shows the way in which sales assistants in one department store negotiated competing pressures from managers (who had become more intent on principles of consumer service), from within the workforce (for example, the fact that 'concessions' staff were able to wear clothes they had chosen themselves while all other staff had to wear uniforms was a source of friction) and from more and more demanding consumers. For many sales staff the increasing emphasis on the consumer had taken place directly at their expense, by providing them with less and less latitude at work. In turn, this diminished their sense of 'ownership' of the store:

> We're us and they're them, yeah. What they don't seem to understand is that we're customers too. We go shopping too. It's almost like they think we're slaves. We don't leave here but go into a little corner where there's beds and we go to sleep there and get up the next morning and come out in the shop . . . There's no idea amongst the customers that we're just at work.
>
> (shop assistant cited in du Gay, 1996: 160)

At the same time, the emphasis on the sovereign consumer has, if anything, made the 'face work' of sales skills more important in many retail organisations, leading to more and more systematic attempts to inculcate such skills into the workforce, a development which, to an extent at least, the workforce often 'resists', for example through 'backstage' irony and mockery (Crang 1995; Crewe and Lowe 1996). Then, second, there are the major organisational changes going on in certain parts of the economy which are leading to greater and greater 'demand pull' (Lowe and Wrigley 1996). In particular, in countries like Britain and the United States, retailers have often become more and more powerful at the expense of producers.[1] In turn, this new relationship means that consumers are more able to directly register their changing preferences on producers which, in turn, have to rapidly adjust their output. For example, Cook and Crang (1996) show how Euro–American consumer preferences for exotic fruits have in turn produced far-reaching effects on producers in the countries that export these fruits (see also the papers in Howes 1996).

Another concern of the third stage is the emphasis on not just the consuming subject but also on the consumer object in the belief that 'taking artefacts, images and performances as quasi-texts is to overlook their most fundamental properties so far as users and witnesses might be concerned' (*Journal of Material Culture* 1996: 8). What exactly is the relationship of the subject and object in consumption? There are a number of views of how this 'in-between' relationship might be framed so as to produce a better sense of the object world. In one of these, which follows the work of Merleau-Ponty, the intention is to formulate a phenomenological knowledge which mobilises notions such as 'the flesh' and a metaphorics of touch in an attempt to capture the intimate, sensual aspect of the subject–object relationship (Game 1991; Grosz 1994). In practice, this means concentrating in particular on the different mediations through which the subject experiences the object, and vice versa. In another view, which follows the work of Benjamin, subject and object are confounded by a tacit, everyday knowledge: 'the tasks facing the perceptual opportunities at turning points in history, cannot be asserted, be solved by . . . contemplative means but only gradually by habit, under the guidance of tactile appropriation' (Taussig 1992: 12). In particular, Benjamin was concerned with understanding the new forms of tactility that were swarming over the visual register as a result of the invention of photography and film. One other view is so-called actor-network theory – here the subject and object are co-produced in heterogeneous networks. Or, as, Callon and Law (1995: 501) put it, 'there isn't a reality on the one hand, and a representation of that reality on the other. Rather, there are chains of translations. Chains of translation of varying lengths. And varying kinds. Chains which link things to texts to things, and things to people, and so on.' Then, in one more view, there is a more general emphasis on 'material culture' which argues that the current dichotomy in Western societies between persons and things is historically and geographically transient and which, by drawing on diverse anthropological and historical evidence, has attempted to move beyond what is still a pervasive humanism.

A third concern has been with the constitution of subjectivity. This concern has been generated by a series of related puzzles that arise out of work which, in the second stage, too easily and unproblematically assigned consumer objects to subjects, in part because of an overdetermination of social categories such as class and status (Mort 1996). This concern might be framed as three different questions about how consumer objects produce subjects. There is, to begin with, the question of what might be termed 'singularity'. How is it that 'the range of specificities that we may inhabit comes together in singularity' (Probyn 1996: 24). Consumer objects clearly have a crucial role in producing the singular person and space.

Then there is the problem of ascribing general meaning to consumer objects, when the meanings are always worked out *performatively*, according to situations that pertain in quite specific spaces and times, situations which are often constrained in terms of the kinds of meanings they make possible or visible. Thus, for example,

> material objects are clearly implicated developmentally in the creation or maintenance of a sense of self. Yet it is hard to argue that the importance of a cuddly toy or comfort blanket for an infant derives from the communally printed image. It is especially difficult to suggest that the reasons why an old person may treasure family photographs is because of the symbolism (if there is any) attached to photograph albums themselves.
>
> (Campbell 1996: 103)

And there is also the problem of 'rationality'. It is still often assumed that consumers make definite choices as a result of discrete actions, even in post-modernist accounts (Campbell 1996). This assumption often underlies even the work of those who realise the problematic nature of this depiction. Thus many commentators smuggle a rational account back in, for example by adopting a distinction between habit and conscious choice (Campbell 1994, 1996).

Thus, what is now transpiring might be called a more practical approach to consumption, one which is based in a notion of everyday consumption practices as 'basically repetitive, intuitive *and* inventive' (Hermes 1993: 497). Such a notion displaces the vocabulary of rationality, choice, and representation by a vocabulary of joint action and embodiment (Thrift 1996). Thus consumption is seen as a practical–moral and contextually specific activity, rather than an intellectualised and abstract system of knowledge, which results from the intersection of numerous actor-networks.

A fourth concern has been with the history of consumption, precisely as a reaction to a standard history of consumption which intertwines it with a history of modernity. As historians have become increasingly interested in consumption so they have overturned a number of the assumptions associated with the standard history. Two corrections have been particularly important. Chronologically there has been enough work now on seventeenth-, eighteenth-

and nineteenth-century consumption in Europe and North America to push back the borders of what can be regarded as 'modern consumption' practices to the early eighteenth century, or even the late seventeenth century. In turn, many apparently novel consumer practices have been shown to have deep historical roots. For example, one of the reasons why department stores come late to Britain is because of the extraordinarily sophisticated history of shops and shopping that predated them (Glennie and Thrift 1996). Second, the sheer variation in ways of consuming has been demonstrated which, combined with anthropological evidence, raises doubts over the possibility of any definitive history of 'modern' consumption. For example, British and North American consumption have been shown to have very different roots involving a much greater degree of standardisation of North American goods from an early point in time (Glennie 1995).

A fifth and continuing concern has been with space and place. In particular, space and place are seen as crucial elements of consumer *identities*. For example, Mort argues that space was not incidental but central to the constitution of a new gay consumer market, and to gay identities:

> Judged by their own standards of entrepreneurship the attempts to forge a new market for young men and their goods in the 1980s were an undisputed success. In the face of the later economic downturn, the commodities which sought to speak to men thorough their gender expanded rather than contracted. Consumer journalism, clothing, toiletries, together with a plethora of other personal objects, continued to display the characteristic imprint of the new formula. Likewise, those zones of the city which provided the settings for the associated consumption rituals proliferated. London, along with most of the other Western metropolitan centres, has a burgeoning homosexual quarter, where commerce, community and sexual politics exist alongside more mainstream practices of city life. There has been no closing down of these projects. The commercial experiments in masculinity appear to have been long-term, not merely transient.
>
> (Mort 1996: 204–5)

Going shopping

Ironically, shopping itself has only rarely been the focus of work in consumption in any of these three stages; commentators on consumption have rarely paid much attention to shopping. Even studies of department stores and shopping malls devote remarkably little attention to the cultural practices of shopping. Instead these practices are subsumed into a more general interest in an overarching activity called consumption. The shopper therefore nearly always figures as a sign for something else. This book takes the opposite tack. Our concern is with what shoppers do and what they understand as 'shopping'.

7

What succour can we take from the available literature? Notwithstanding the general neglect of shopping, six main accounts can be identified.[2]

The first of these might be termed 'the commodity regnant'. In this account, shopping figures as an index of the imminent or actual decay of Western civilisation resulting from commodification. In their different ways, both Baudrillard and Bauman see shopping as a form of seduction by the commodity calculus; empty games for empty people; the *flâneur* turned into just another shopper:

> The right to look gratuitously was to be the flâneur's, tomorrow's *customer's*, reward. Pleasurable display, fascinating view, the enticing game of shapes and colours . . . bought through the seduction of the flâneur; the flâneur, through seduction, was transformed into the consumer. In the process, the miraculous avator of the commodity into the shopper is accomplished. At the end of the day the dividing line has been blurred. It is no more clear what (who) is the object of consumption, who (what) is the consumer.
>
> (Bauman, 1993: 173–4)

A less hysterical and more innovative version of this account is provided by Bowlby (1993). For her, shopping is both the first and last sign of the expansion of the commercial sphere into everyday life: 'all the world's a showroom, every man or woman an advertisement for himself or herself, aiming to "impress" and please his or her consumers' (Bowlby 1993: 95). Thus, Bowlby provides a blow-by-blow account of the drama of the sale, and questions whether discourses of the self have been infected by the drama:

> There is an intimate connection, institutionally and intellectually, between psychology and marketing during the first forty years of this century and beyond. As psychology became separated off from philosophy on the one hand and neurology on the other as an independent discipline, the primary questions it was concerned with were often identical to those that preoccupied advertisers who wanted to know how people acted in order to know what would get them to buy.
>
> (Bowlby 1993: 114)

Even Freud's writing shows traces of this motivation. Ironically, given Freud's disdain for consumer society, especially in the shape of the United States,[3] 'I shop, therefore I am' seems to be a slogan that is less far from his concerns than might at first be thought:

> Freud's own writings, looked at through the lenses of consumer psychology, might seem to be offering another version of the modern marketing mind. It is a commonplace to talk about the 'economic' model in Freud, but this is never, as far as I have seen, put into

relation with either the economics of his time or the psychological preoccupations of that economics in the area of marketing.

(Bowlby 1993: 114)

For Bowlby, Freud's writings contain both a logic of the rational psychic operator, a *homo economicus* dealing in a calculus of value, cost and saving and also a logic of comparison shopping and impulse buying:

> There is a drama of attractions and interests, desires and choices, in which minds are forever seeking and forever failing to capture the object that will satisfy their longing once and for all. The choices and wishes of love appear in the same linguistic guise as those of consumption: one of the words used in relation to object-choice, for instance, is the verb *auszeichnen*, which is primarily used for purchasing something in a shop.

(Bowlby 1993: 115)

In the second account, shopping is also about the commercial sphere and commercial capitalism, but, in contrast to the account of an all-consuming capitalism, it attends to the diversity of forms of capitalism. The 'new retail geography' finds a messier and more disparate field of action which consists of a spectrum of retailing firms and markets, operating through distinctive geographies of shopping malls, department stores, supermarkets, discount warehouses, corner shops and so on (Wrigley 1991, 1992; Marsden and Wrigley 1996; Wrigley and Lowe 1996). In turn, these geographies offer different kinds of shopping experience and demand different kinds of shopping knowledge. New forms, markets and geographies of shopping are constantly being formed. For example, Lowe and Wrigley (1996) point to the rise of a whole series of new forms of shopping in the last 15 years, including the 'captured market' which consists of sites intended for other purposes which have become shopping opportunities (e.g. airports, railway stations, petrol service stations, hospitals), 'taking consumption to the consumer', as in carefully targeted shopping catalogues and lifestyles magazines, and the 'leisured consumer', which consists of shops centred on leisure themes (e.g. the Disney and Warner Brothers stores, football and rugby club stores). The point about this literature is that it depicts a world of constant tension between producers, retailers and consumers, which is kept in balance by specific regulatory regimes which themselves incorporate particular notions ('rights to produce and consume') of the apposite relationships between producers, retailers and consumers. More than this, the literature ties shopping back to production and in doing so, reminds us of the ethics of shopping.

A third account figures shopping as part of a refocusing of connections between the commodity and post-traditional identity which is incorporated in the notion of 'lifestyle' (Chaney 1996; Shields 1992a). Consumers gather around objects which define their identity and become centrepieces of particular routines of sociability:

> Lifestyles are routinised practices, the routines incorporated into habits of dress, entry, modes of acting and favoured milieus of encountering others; but the routines are reflexively open to change in the light of the mobile nature of self-identity. Each of the small decisions a person makes every day . . . contributes to such routines.
>
> (Giddens 1991: 81)

Lifestyle therefore differs from traditional status orders, as well as structural divisions (class, family, gender, ethnicity) and for two reasons (Slater 1997). First, it tends to stress a powerful cultural pattern made up of signs, representations and media. Second, and relatedly, it is inherently unstable since, in a sense, it is an extension of consumer choice to mode of life:

> lifestyle groupings and patterns do not reflect communities with well-policed social gates, with obligations to long-term commitment or to extensive social learning processes. Moreover, lifestyle groups are 'elective communities', memberships which we choose rather than have ascribed or allocated to us. Social membership is reduced to identities one puts on and turns off at whim, a flânerie which moves us beyond even the solidarity of subculture to 'the supermarket of style'.
>
> (Slater 1997: 88).

Unlike its subject, the lifestyles literature on consumption can have a rather mechanical feel, often coming perilously close to the kind of consumer categorisations that still bedevil the literature. But it does not have to, as the work of Shields (1992a) attests. Shields constantly stresses the ambivalence that lies at the heart of lifestyle as concept and social practice.

In the fourth account, contemporary shopping is a reflection of a new and more democratic definition of heritage and further evidence of an expanding historical culture. Shopping becomes a part of this expansion of popular memory, as shoppers become Clio's under-labourers and goods become clues and signposts to the past. This expansion of historical culture into shopping can be indexed in a number of ways. There is, first of all, the new visibility of shopping in representations of the material past:

> In pictorial histories (such as those reproduced from old postcards . . .) pride of place is given to the high street scene. In the mock-ups and pin-ups of the 'traditional' village, the general stores – or village post office – occupy the symbolic spaces once given to the parish church. 'Period' shopping is a leading attraction at the open-air

museums and theme parks. At the York Castle Museum, a pioneer in this field, a cobbled Victorian street has been reconstructed, with complete shop fronts reused as architectural salvage ... Shops of a more recent vintage are also the centre piece of the open-air and industrial museums. At Beamish visitors are transported from place to place in enamel-encrusted trams. Shop assistants in 1920s costume serve them.

(Samuel 1994: 160)

This cultural revisionism extends even to the shopkeeper:

He is no longer the obsequious figure of nineteenth century carica-ture, fawning on the carriage trade, nor yet a melancholy Mr Polly, teetering on the edge of bankruptcy ... nor yet the vulgar commer-cial of Matthew Arnold's *Culture and Anarchy* but rather, like the old-fashioned draper, an emblem of 'knowledgeable and friendly service'. In the book of sepia photographs, he is a figure of authority, flanked by respectful assistants, and backed by mountains of produce. In the trade catalogues, often reproduced in sepia in recent years, shopkeepers figure, as in Whiteleys of Queensway, as 'universal providers'. In oral history's childhood memories they are fondly remembered as the purveyors of broken biscuits and spotted fruits.

This new version of the national past is not only more democratic than earlier ones but also more feminine and domestic. It privileges the private over the public sphere, and sees people as consumers rather than – or as well as – producers.

(Samuel 1994: 161)

A second means of expansion has been the historicizing of consumer goods. Samuel provides a vast inventory of 'historic' fare, ranging from 'traditional' foods through toiletries, china, clothing, music, housing, and even lighting. He finds, for example, one glass factory in Wakefield that is using 'state of the art modern technology to produce traditional old-style uplighters and shades' (Samuel 1994: 103). Then third, there is the expansion of sites of memory. There are the carefully manicured centres of historic towns such as Bath, Chester, and Oxford. There are period shopping streets and shopping precincts, full of shops with small paned windows and carefully calligraphed hanging signs. There are the retail chains selling historic goods, such as Past Times, which makes a fetish of the facsimile.[4] And then there are the ubiquitous 'historic' touches to be found in so many shopping areas: Victorian street furni-ture, mock street lighting and so on.

A fifth account considers shopping predominantly in terms of gender and sexuality. All manner of researchers have noted that gender and sexuality are important aspects of shopping. In terms of gender, the first point to make is the degree to which shopping is equated with women: 'the vast majority of the world's shoppers have been women' (Reekie 1993: xi). Further, from the middle of the nineteenth century, women came to predominate on many shop floors as shop assistants. However, men also have a shopping history. In the nineteenth century, men were a rare find in many shops. For example:

> there is evidence which indicates how women continued to exclude men from the feminine terrain of the department store. The women workers in Marshall Field's were noted for their scornful giggling of men who accompanied their wives in the store. These 'Molly Husbands' suffered both the ridicule of breaking late nineteenth century gender stereotypes and the subtle-tactics of both women customers and workers defending their 'space' . . .
>
> Many stores appear to have colluded in this process of male exclusion by providing 'Men's shops' with separate access. One Blackpool store even went so far as to provide a men's room complete with newspapers, free cigars and coffee.
>
> (Lancaster 1995: 182)

Sexuality is an equally important element of this account. There are modes of shopping in which sexuality is a clear element (see, for example, Nead 1997, on a London street famed for pornography), but more generally, shops can be seen to have been responsible for the creation of a series of sexual cultures based around new practices of shopping. For example, Reekie writes of the department store that its significance:

> lies in its implicit management of male and female bodies and sexual relations. New methods of display, advertising, customer surveillance and departmental organisation introduced after the turn of the century ascribed increasingly explicit cultural meanings to male and female bodies. Mass marketing methods also articulated with greater clarity differences between the sexes.
>
> The new sales techniques and knowledges provided unambiguous and frequently tangible models of appropriate modes of heterosocial conduct. Like the superficially rational but deeply sexualised environment of the modern office, department store culture was saturated with the signs of sex. The department store was an institution of everyday culture which touched the lives of every woman who shopped there; hence, its potential to shape popular values and ideas about sex and gender was invaluable. The department store in effect created modern manhood and womanhood and, as a heterosocial space, crucially influenced the ways in which women and men viewed and related to each other as sexes.
>
> (Reekie 1993: xiv)

Similarly, at a later date, shopping has clearly had an important influence on the growth of new sexual relations. Most particularly, men have become more closely involved in shopping:

> the gender of shopping has changed dramatically in the post-war period. The days of the 'Man's Shop', with its discreet separate entrance, have long passed. In the modern post-war period, shop assistants at Brown's told the Mass Observation investigators that they had witnessed an increase in the number of men accompanying their wives on shopping expeditions. The Mintel Survey of 1978, which claimed that ten per cent of all adult men visited a department store at least once a month indicates that male prejudice about shopping was beginning to break down. A visit to any high street or mall on a Saturday visibly demonstrates the growth of male participation in the experience of shopping. Men still insist that the purchase of 'technical' products, such as home electrical goods requires their 'expertise'. Men are also reported to be more prone than women to impulse buying.
>
> (Lancaster 1995: 202)

This involvement of men in shopping was boosted in the 1980s by the development of what Mort (1996: 9) calls 'a commercial language capable of speaking to young men' by the advertising, marketing and retailing industries. In turn, this language was imported into a number of sexual cultures, including those of the burgeoning gay communities. Commodities and the practices of shopping featured as 'recognisable signposts' (Mort 1996: 11) through which these new sexual cultures could be enacted.

The sixth account tells of shopping through the eyes of the shop-worker. Shop-workers have tended to be neglected in the literature. For example, Lancaster (1996) could find only four minor studies in Britain on department store labour in the industrial relations literature, although the situation was better with regard to American and French department stores. However, the paucity of material is, to an extent, made up for by the quality of what does exist. There have, for example, been classic studies of department store labour in the 1950s (Woodward 1960) and of the John Lewis Partnership in the 1960s (Flanders et al. 1968). More recently, studies like those of du Gay (1996) have shown the influence on the shopfloor of management theory. Nearly every study conceives of shopping as a constant battle between workforce and consumers; 'over a century, from the tabbies of the late nineteenth century to the foragers in the present-day bargain basement customers and managers have pitted their wits against each other' (Lancaster 1995: 169). In particular, wits have been sharpened by shoplifting, which to go back to the previous account, had, and has, a clear gender dimension (Abelson 1989):

the sexual division of department store life was closely associated with another such phenomenon, shoplifting, an activity that taxed the minds of many Victorian commentators. Pilfering from shops is as old as retailing; but what captured the popular imagination in the second half of the nineteenth century was that much of this crime, especially in larger stores, was committed by middle class women.

(Lancaster 1995: 184)

What each of these six accounts show is that shopping is a complex social activity which needs to be treated as such. As Lancaster puts it:

we need to keep in mind the wide variety of shopping experiences. To take the department store as an example, on any day we confront the bargain-hunter in the basement, the parents buying their children's school clothing, a couple buying white appliances or a television after lengthy research through copies of *Which?*, the browser gazing at expensive designer label clothes, women meeting socially in the coffee shop, the children being brought toys ... the list is almost endless and warns against simplistic generations.

(Lancaster 1995: 169)

In what follows we attempt to outline some of the key principles that underlie our approach, distilled out of the literatures that we have already reviewed, without aiming, at this point, for a hard and fast theoretical stance.

Towards a theoretical framework

First, then, 'shopping' is a term we use to denote a network of activity of which the actual point of purchase of a commodity is but a small part; 'shopping for goods remains a social activity built around social exchange as well as simple commodity exchange' (Shields 1992b: 102). Thus, even at a site of purchase like a shopping mall 'there are pro-forma greetings and salutations, a banal Goffmanesque interaction between scripted roles of shop assistant and shopper – a shopper who may or may not become a purchaser' (Shields 1992b: 102). Then, the sale of any good nearly always requires an extensive infrastructure of relations of production, distribution and marketing which cannot be excluded from it. It also depends upon the intersection of networks of the social with different obligations and objectives which can produce very different stances towards the same goods. Finally, it always involves a process of 'translation' as the product is taken home and sieved through the varied cultural milieu which give these objects their social meaning (Miller 1987). Thus:

as a consequence of consumption work, consumption cannot simply be reduced to the nature of the commodity and the consumer is more

14

than simply the process by which the commodity is obtained. Rather, through the contribution of intrapsychic, biographical, family, gender and cultural forces, a person-object relation is regulated which in turn gives rise to identities, understanding and everyday practices.

(Lunt and Livingstone 1992: 85)

Second, shopping is normally an everyday activity. It is not therefore constantly reflected upon. Many shopping skills are practical, learnt, literally, at a mother's, or father's, knee through performative explorations involving a mixture of mimesis and being a part of conversations. Such skills are therefore always joint, embodied, and practical–moral in character (Thrift 1996). They are also always attuned to the situation at hand and involve a skilful copying, a readiness to deal 'appropriately' with people and things. Examples of these skills are legion. They range from knowing how to push a defective supermarket trolley through the practical calculative skills Jean Lave (Lave 1986; Lave and Wenger 1991) has so interestingly documented amongst supermarket shoppers to 'having a nose for a bargain' and being able to deal with an intimidating shop assistant. And these skills, as we note in a subsequent chapter, are highly valued.

But, third, this is not to say that shopping is utterly unreflective. In particular, it intersects with all kinds of discourses which people participate in to a greater or lesser extent. These discourses which are as much passed on by people through conversation as they are passed down from the media involve, most particularly, the constant worrying away at dilemmas which are the result of shared bits and pieces of social knowledge which are in conflict (Billig et al. 1988). Thus:

the rhetorical approach does not start by considering individual motivations or individual information processing. It starts from the assumption that knowledge is socially shared and that common sense contains conflicting, indeed dissonant, themes. It is not neatly systematised in a way that permits the individual who has dutifully accepted society's values to generate automatically all necessary thoughts, actions and argumentative discourse. Instead, common sense provides the individual with the seeds for contrary themes, which can conflict dramatically in dilemmatic situations. Because these are seeds, not flowers, all is not fully systematised. Contained within the conflicting general principles are many different possibilities, which may on occasion give rise to argument and debate. Rather than applying their systems unthinkingly, people must also deliberate and argue about which seeds need planting at which times in order to develop into flowers. And when people so debate or argue, then living has a dilemmatic quality.

(Billig et al. 1988: 20)

The dilemmatic quality of discourses is clear in the shoppers' accounts given in subsequent chapters. For example, 'nature' represents a series of key dilemmas for shoppers. Nature represents a simpler, more attractive reality. But that representation can be brought (or should it be bought) into question by shopping. Price (see also Smith 1996) provides a good example of this disruption in her description of the US retail chain, The Nature Company, which, as an anchor of many malls, is targeted at middle class consumers:

> The store sounds like fun, and it is. So why do I feel ambivalent? . . . Why does the store 'feel false' to some of its patrons? Well, to begin with, the Nature Company is not nature. And among the set of meanings we've attached to the natural world, perhaps to most overarching and powerful is that nature is *not* a shifting set of human meanings. It's tangible, secure, rock like, stable, self-evident, definable, real. In a word, it's natural. Not that we don't know or acknowledge that nature means definable things to us, like 'solitude' or 'relaxation'. But the meanings themselves seem universal, indelible, indigenous to the rocks and trees themselves. And nature, we tend to assume, is for everyone, or should be. 'Nature' is not constructed, like a movie, to tell a story that appeals to a definable audience in a certain time and place.
>
> Ordinarily if you buy your pruning shears at the hardware store, or your bird guide at a bookstore, these convictions don't face any serious threat. In fact, my friend testified that a trip to Ace Hardware for gardening tools would not feel like an 'inauthentic' experience, nor would she feel 'manipulated'. In the garden, too, where you're surrounded by what is undeniably real and tangible about nature, nature feels seemingly natural. But, here, where the Nature Company has brought together thousands of nature-oriented products, the boundaries we've drawn around 'nature' begin to become visible. If you compile the complete pool of meanings, and stack and shelve them all together in one room in a mall, they begin to look like meanings. And in this upscale venue, practically neighbours with Emporio Armani, whose meanings they are becomes an almost palpable question. Few of us will respond with, 'Aha, so the meaning of nature is not so self-evident or universal after all'. The response, I think, is closer to 'um, wow'. The store invites us in, but plants the vague suspicion that nature is a very human, historically shifting idea – not precisely what most of us are shopping for.
>
> The Nature Company is engaged in a highly tricky pursuit. It's marketing a product – middle class meanings of nature – to target consumers who tend to question the product's existence. The company also markets 'authenticity', 'uniqueness' and 'simplicity' in the extravagant maw of South Coast plaza. It's a lucrative business if you can do it, but the very meaning that the 'Nature' in 'the Nature Company'

immediately calls to mind – the antimodern associations that are the company's real commodity provoke many nature lovers to doubt the most basic features of the enterprise. The Nature Company, tapping flawlessly into the market for anatomically correct inflatable penguins, and the perfect place to go to encounter what nature means to America's 'affluent middle aged' in the late twentieth century breeds some distrust among its clientele.

(Price 1995: 190–1)

Fourth, shopping is about social relations. We take it as axiomatic that we live through others, in joint action with them. It is no surprise, then, that shopping is as often about others as it is about self (and even when it is about self, it is often still about others). In our society, the chief of these others is the family, a structure of affinity that few of us escape: as Adam Phillips (1994: 39) puts it: 'if sex is the way out of the family, falling in love is the way back, the one-way ticket that is always a return'. It is no surprise, then, as we show in chapter 5, that shopping is so often about the moral economy of the family. Buying goods is often as much about others in the family as it is about the shopper, especially for women. Of course, family is not only a support but also a constraint, and many of our shoppers feel this dilemma constantly. They cannot live without family affinity but family affinity also brings tiresome obligations, both on an everyday basis and most especially in the case of significant consumer festivals associated with the family – Christmas, birthdays, Mother's Day, Father's Day and so on. Saddest of all are the cases of persons who try to repress time by fulfilling obligations that other family members no longer reciprocate: here the buying of goods becomes a kind of emotional cargo cult.

Some commentators have suggested that, besieged by consumer capitalism, the family is on the wane as a network of social obligations and affinities. The evidence, in fact, seems to suggest the opposite. As Pahl (1995: 179) puts it: 'as long as we do not make the silly mistake of conflating household with family, it is arguable that the family and family relations are now, in the 1990s, stronger than at any time in the past'.[5] But what is clear is that the family is being paralleled as a network of social obligation and affinity by other social networks based on friendship (Allan 1996; Giddens 1991).

These 'families we choose' (Weston 1991), based on criteria such as sexuality, are a significant element of modern societies. Thus 'choosing certainty' is bolstered, even underpinned, by goods which underline the value of interpersonal relationships (as in the 'Forever Friends' line of greetings cards). In so doing, one might argue, as have Finch (1989) and Strathern (1996), that these families we choose are, in a sense, simply reclaiming traditional familial virtues.

What seems clear is that, if anything, social relations have become more rather than less dense, and maintaining these relations in turn demands constant and considerable consumer expenditure. Thus, on the question of whether people in Britain lead socially isolated lives one commentator provides a very clear answer, that

> most people do not. Most of us have personal networks which contain large numbers of others and within which a smaller number are particularly important to us. Some of these important relationships are with kin and some with non-kin. The patterns here do vary depending on a wide range of circumstances. And equally, some people do lead quite restricted lives and are socially isolated. The point, though, is that this is a relatively unusual rather than the normal state of affairs in contemporary society.
>
> (Allan 1996: 129)

Fifth, shopping is about the goods themselves. Consumer objects are not neutral participants in the practices of shopping. They both embody social relations and extend them in new directions. They hail subjects in different ways, as we will see in the use of certain forms of clothing favoured by younger people. They are objects with which people are more or less acquainted, according to their pattern of use and position in a moral universe. And they have their own biographies (Appadurai 1986).

Such a depiction of objects as active participants in consumption does not mean having truck with what Morris (1988) calls 'commodity boudoir talk', in which the commodity is refigured as responsible for a seductively fallen state – commodities call out to each consumer pulling them into a tainted fantasy world by offering a false intimacy. As Morris, alive to the sexual overtones of this metaphor, makes clear:

> First, one might ask, what is the sound of an intimate and ad hominem address from a raincoat at *Big W*? Where is the secretive isolation of the thongs [sandals] in a pile at *Super K*? The commodities in a discount house boast no halo, no aura. On the contrary, they promote a lived aesthetic of the serial, the mechanic, the mass-reproduced; as one pair of thongs wears out, it is replaced by an identical pair, the same sweat shirt is bought in four different colours, or two different and two the same; a macramé planter defies all middle class whole-earth naturalness connotations in its dyes of lurid chemical mustard and killer neon pink. Secondly commodity boudoir talk gathers up into the style and class-specific image of the elite courtesan a number of different relations women and men may invent both to actual commodities, the activity of combining them and, above all, to the changing discursive frames (like shopping centres) that invest the practice of buying, trafficking with and using commodities with their variable local meanings.
>
> (Morris 1988: 221)

18

Rather, it suggests that objects are social relations made durable. Even simple objects can produce particular relations. For example, Latour (1992) provides the example of an automatic door-closer (in part, introduced to obviate the need for human labour like a porter) which requires swift reflexes and movement on the part of human users, a bodily stance which makes life difficult for small children, the disabled and the elderly. More generally, we can point to the increasing density and complexity of consumer objects, our more 'charitable' (Collins 1990) stance to them and an increasing likelihood of imagining ourselves through these objects.

If there is a consensus concerning the place of consumer objects it is therefore that they are not a part of a separate 'material' realm. Rather, to use the terminology of actor-network theory, they co-produce reality. As Miller puts it in discussing technology:

> What is clear . . . is that it is not some prior moral culture into which technology is placed, but that technology and life-style are the form through which the ideal of morality is itself constructed. If it is ever abstracted as morality this is done on the basis of the lived relationships, where people discover what their morality is: it is the interaction with the technology which makes the morality as much as the other way round.
>
> (Miller 1992: 225)

Sixth, and finally, shopping is about place and identity. One of the key themes of this book is that particular parameters of identity such as the family, class, ethnicity, and gender are reconstituted by shopping sites through the addition of particular distinctions which emerge from the experience of these spaces. This principle is sufficiently important that we address it in more detail below.

Theorising place and identity

'Identity' has become one of the keywords of the 1990s, denoting both the social recognition of difference and culturally constructed notions of the Other. In particular, the notion has been associated with the rise of various forms of 'identity politics' as older, rigidly defined, class-based identities have given way to a variety of other sources of identity and their associated social movements.

The shifting nature of personal identities in late modernity has been debated at length (Giddens 1991; Beck 1992; S. Hall 1992a; Lash and Friedman 1992). For Giddens and Beck, in particular, contemporary identities can be theorised as a reflexive project, shaped by the institutions of late modernity and sustained through narratives of self-identity that are continually monitored and constantly revised. Such an approach is, in part, consistent with our own. But it has to be seen as applying chiefly to what might be called the vanguard of social development where there may be more focus upon the individual. Our empirical work, in contrast, finds that identities are still largely relational. In particular, and in line with the authoritative work of writers such as Finch (1986; Finch and Mason 1990) and Allan (1996), we find that the family still remains as the core context for self-development.

Where our appeal is more in accord with current debates is in its emphasis on identity as a social process that shifts according to social context. Thus, throughout the project, we have explored the way that identities can be expressed in relation to particular places and particular material goods. In other words, we approach identity as a *discursively constituted* social relation, articulated through narratives of the self and accessed empirically through focus group research and ethnographic methods.[6] Or as Somers puts it:

> Narrative identities are constituted by a person's temporally and spatially variable place in culturally constructed stories composed of (breakable) rules, (variable) practices, binding (and unbinding) institutions, and the multiple plots of family, nation, or economic life. Most importantly, however, narratives are not incorporated into the self in any direct way; rather, they are mediated through the enormous spectrum of social and political institutions and practices that constitute our social world.[7]
>
> (Somers 1994: 635)

Rather than seeing identity as fixed and singular, it has become common to think of identities as plural and dynamic. For, as Frank Mort argues in his study of the consumption styles of fashionable young men in London:

> We are not in any simple sense 'black' or 'gay' or 'upwardly mobile'. Rather we carry a bewildering range of different, and at times conflicting, identities around with us in our heads at the same time. There is a continual smudging of personas and lifestyles, depending where we are (at work, on the high street) and the spaces we are moving between.
>
> (Mort 1989: 169)

Furthermore, although individuals as identities represent specific trajectories, these are related to normative and socialised categories which are objectified by their sites of objectification. So, in Mort's case, what it means to be gay becomes centred upon what the area of Soho in London has come to represent. Similarly, in our work, the shopping centres themselves become a form through which the meaning of class is understood and taken up in individual identity formation.

Thus, as the passage from Mort suggests, theories of identity are increasingly articulated in relation to particular spaces and places, both in the metaphorical sense of boundaries, domains and diasporas (Rutherford 1990;

Carter et al. 1993) and also in relation to specific spaces and places (Keith and Pile 1993; Pile and Thrift 1995). In one sense, this is ironic since traditional attachments to place may no longer offer clear support to our shifting identities: 'The presumed certainties of cultural identity, firmly located in particular places which housed stable cohesive communities of shared tradition and perspective, though never a reality for some, [have been] increasingly disrupted for all' (Carter et al. 1993: vii).

Hence the need for new theories of space and place which are more consistent with the increasingly hybrid character of contemporary identities, blending different sources and traditions from different locales. One such is Doreen Massey's (1994) argument for a 'progressive' or global sense of place. For Massey, rather than searching for a lost 'authenticity' based on nostalgia and false memories, places should be understood as a distinctive articulation of social relations from the global to the local. According to Massey, places are not bounded areas but porous networks of social relations, constructed through the specificity of their interaction with other places. Massey gives the example of Kilburn High Road, a pretty ordinary street in north-west London which is linked through a web of culturally inflected social, political and economic relations to places as far apart as Ireland and the Indian sub-continent. Kilburn may have a character of its own, Massey argues, 'but it is absolutely not a seamless, coherent identity, as single sense of place which everyone shares' (1994: 153). Thus, Massey provides a means of reconciling the emphasis upon individuals as reflexive creators of identity and spaces as normative objectifications of identity through a sense of identity as a practice in which the ambiguities and pluralism of space allow for related but varied forms of appropriation by individuals, households and larger social fractions.

These notions, which we employ in our study of Brent Cross and Wood Green, clearly parallel contemporary theories of identity such as those associated with the works of Paul Gilroy, bell hooks and Homi Bhabha. Summarising this literature, Edward Said argues:

> No one today is purely one thing. Labels like Indian, or woman, or Muslim, or American are now more than starting points, which if followed into actual experience for only a moment are quickly left behind. Imperialism consolidated the mixture of cultures and identities on a global scale. But its worst and most paradoxical gift was to allow people to believe that they were only, mainly exclusively, White, or black, or Western, or Oriental.
>
> (Said 1993: 407–8)

Traces of such global influences are present throughout the following chapters in relation to the increasingly global sourcing of commodities, for example, and in terms of how people feel about their attachments to the neighbourhood and the nation.

The array of identities with which the modern consumer is currently faced may give rise to a sense of risk and uncertainty (Beck 1992; Warde 1994). But these risks can be 'managed' by ordering them into coherent 'lifestyles', subject to various forms of social regulation. As Judith Butler (1990: 135) argues in relation to the permeable boundaries of gender, our identities may be a performative creation but they are made culturally intelligible through regulatory grids such as those associated with an idealised and compulsory heterosexuality. In this sense, identities can be thought of as a 'regulatory fiction'.

And yet most consumption studies, based firmly in the quantitative tradition of marked research, continue to adhere to a simple typology where identities are classified according to a binary, either/or, system. In a recent *Retail Intelligence* report, for example, the consumer research organisation MINTEL (1994) classified shoppers into five mutually exclusive categories, which break down by gender in a highly stereotypical pattern (see Table 1.1).

According to such typologies, men are simply more 'reluctant' or 'obstinate' shoppers than women. There is no understanding of the circumstances in which such stereotyped views might not apply or how particular individuals may be 'purposeful' in relation to some kinds of purchases while being 'addicted' to other kinds of shopping.

While such categorisations have understandably lost favour within contemporary social theory because of their implication that identity is singular and static, similar schemes are still widely used in social science research on consumption. In reviewing the literature on the 'modern consumer', for example, O'Brien and Harris (1991: 120–1) provide a litany of such classifications: status seekers, swingers, conservatives, rational men, inner directed men and hedonists; pragmatists, bargain-hunters, satisfied shoppers and shopping trippers; purposive, time-pressured, fun and experimental shopping. Even some of the most sophisticated of recent studies, such as Lunt and Livingstone's *Mass Consumption and Personal Identity* (1992), utilise a simplistic schema, classifying their respondents into five mutually exclusive categories: alternative, routine, leisure, careful and thrifty shoppers.

Such categorical approaches may be a necessary prelude to statistical analysis but they bear little relation to the shifting and complex nature of social identities

Table 1.1 Market research classification of shoppers

	Male	*Female*
Per cent classified as:		
addicted	8	22
happy	13	24
purposeful	36	34
reluctant	24	13
obstinate	18	8

Source: MINTEL, *Retail Intelligence* report 1994, figure 52.

where we rarely think of ourselves as one thing at all times and in all places. Judith Williamson provides a striking anecdote with which to combat such simplistic ideas, admirably expressing the complex relationship between consumption and identity:

> When I rummage through my wardrobe in the morning I am not merely faced with a choice of what to wear. I am faced with a choice of images: the difference between a smart suit and pair of overalls, a leather skirt and a cotton dress, is not just one of fabric and style, but one of identity. You know perfectly well that you will be seen differently for the whole day, depending on what you put on; you will appear as a particular kind of woman with one particular identity *which excludes others*. The black leather skirt rather rules out girlish innocence, oily overalls tend to exclude sophistication, ditto smart suit and radical feminism. Often I have wished I could put them all on together, or appear simultaneously in every possible outfit, just to say, How dare you think any one of these is *me*. But also, See, I can be all of them.
>
> (Williamson 1986: 91)

Theorising consumption as a social process rather than as a single, isolated moment of exchange leads to new ways of theorising identity. Drawing on the work of Pierre Bourdieu (1984), for example, Mary Douglas (1996) suggests that modern identities are constituted through our relationship with the symbolic world of consumption rather than through a direct relationship with the material world. Focusing specifically on 'protest shopping', she finds much of the current literature on 'consumer choice' in market research and social psychology insulting to the intelligence of the consumer. The rationality of the consumer, according to Douglas, emerges from seeing consumption as a choice not just between different *kinds of goods* but between *kinds of relationship*. Douglas applies this argument to an analysis of 'mindless consumerism', where goods are allegedly desired for their own sake, for purposes of self-indulgence or personal display. By showing that the ownership and display of such goods is part of a consistent social pattern, she argues, we can demonstrate how people's tastes and preferences are clearly politicised. The basic choice, according to Douglas, 'is not between kinds of goods, but between kinds of society, and, for the interim, between the kinds of position in society that are available to use as we line up in the debate about transforming society' (1996: 112).

Our own argument is not dissimilar, though rather less grand in its ambition. It is applied at a more domestic scale, in terms of the investment in *social relationships* that takes place during the apparently mundane work of shopping. Instead of reducing identity to a series of 'cultural preferences and lifestyles' – in Hobsbawm's (1996) rather dismissive phrase – we have approached identities as multiple and contested, discursively constituted through narratives of the self, constructed in relation to socially significant others and articulated through relations with particular people, places and material goods. It follows that rather than simply inferring people's identities from the purchases they make, our main interest is in how they narrate their identities, drawing on a relatively limited repertoire of available images and representations. Thus, we do not seek to describe the purchasing habits of different social (family, class, ethnic or gender) groups. Rather, we are interested in the way that narrative identities are constructed by these different groups and in the different discourses on which people draw as they relate to particular types of goods in particular kinds of places.

Shopping malls

So how are identity and place bound together in the practices of shopping? There is, first of all, the obvious point that shopping takes place at a wide variety of different sites, each of which represents often quite different kinds of shopping experience and resources for identification: high streets, super-markets, shopping malls, airports,[8] petrol/service stations, motorway service stations, factory outlet villages, retail warehouses, garden centres, shops in stately homes, museums and art galleries, charity shops, second-hand shops . . . the list goes on. Second, even sites which may appear superficially similar may prove to have quite different practices of use and valuation which, in turn, appeal to certain forms of identification rather than others.

Shopping malls, on which we concentrate in this book, is a case in point. Yet malls are not always seen in this way. There are currently at least three competing interpretations. In the first of these malls are often elevated to the status of bridgeheads of an all-conquering capitalism. They are homogeneous consumer machines, creating consumer homogeneity. Baudrillard's work is perhaps the most obvious early example of this apocalyptic tendency. For him, in the mall, the consumer is caught up 'in a calculus of objects, which is quite different from the frenzy of purchasing and possessions which arises from the simple profusion of commodities' (Baudrillard 1988: 31). This is because the basic concept of the mall is 'ambience' rather than the encoding of factors which are meant to have specific psychological effects on a shopper's choices. As Lunt and Livingstone put it in their explication of Baudrillard's views:

> the shopping mall affords the opportunity of participation in the currency of modern society: exchange (implying a contrast with the structure of Athenian democracy where the currency of partici-pation was argument). The shopping mall is a public forum – the site of participation in late capitalist society as formulated through commoditization. The consumer culture is a new form of the manip-ulation of the ordinary person by the exchange system. When the family goes to the shopping mall together at the weekend, the mall

provides a form of leisure, of structuring time, and a site for constructing family relations of gender and generation. Our identities and experiences are produced by the experience of participation in the cultural form of late capitalism – the shopping mall.

(Lunt and Livingstone 1992: 21)

For Baudrillard, then, a consumer apocalypse is nigh:

We have reached the point where 'consumption' has grasped the whole of life, where all activities are squeezed in the same combinatorial mode; where the schedule of gratification is outlined in advance, one hour at a time; and where the 'environment' is complete, completely climatised, finished and culturised . . . work, leisure, nature, and culture all previously dispersed, separate, and more or less irreducible entities that produced anxiety and complexity in our real life, and in our 'anarchic and archaic' cities, have finally become mixed, massaged, climate-controlled, and domesticated into the single activity of perpetual shopping.

(Baudrillard 1988: 33–4)

This kind of apocalyptic talk still periodically reoccurs in the literature. For example, Bauman seems convinced that the coming of malls marks a kind of postmodern *Independence Day*:

'Malls' in its original meaning refers to the tracts for strolling. Now most of the malls are *shopping* malls, tracts to stroll while you shop and to shop in while you stroll. The merchandisers sniffed out the attraction and seductive power of strollers' habits and set about moulding them into life. Parisian arcades have been promoted retro-spectively to the bridge heads of times to come: the postmodern islands in the modern sea. Shopping malls make the world (or the carefully walled-off, electronically monitored and closely guarded part of it) safe for life-as-strolling. Or, rather, shopping malls are the worlds made by the bespoke designers to the measure of the stroller. The sites of mis-meetings, of encounters guaranteed to be episodic, of the present prised off from the past and the future, of surfaces glossing over surfaces. In these worlds, every stroller may image himself [sic] to be a director, though all strollers are the objects of direction. That direction is, as their own used to be, unobstructive and invisible (though, unlike theirs, seldom inconsequential), so that baits feel like desires, pressures like intentions, seduction like decision-making; in the shopping malls, in life as shopping-to-stroll and strolling-to-shop, dependence dissolves in freedom, and freedom seeks dependence.

(Bauman 1996: 27)

This kind of negative account is carried through in a milder but ultimately no less condemnatory form in much other work on shopping malls (for a review, see Jackson and Thrift 1995). For example, Goss (1993) uses a language of 'plots', 'lures' and 'decoys' to describe what he sees, using de Certeau's (1984) terminology as the 'strategic' space of the malls, a space that seems to have previous little room for de Certeau's 'tactics'; where 'radical soon becomes radical chic' (Goss 1993: 41). The mall is an instrumental landscape, designed by the 'captains of consciousness' to bend consumers to its will. The tale, in other words, is of a glitzy facade behind which lurks a grimmer, exclusionary reality: 'the alienation of commodity consumption is concealed by the mask of carnival, the patina of nostalgia, and the ironic essences of elsewhere' (Goss 1993: 40).

In the second account, the mall becomes the opposite of the apocalyptic account: now it is a polysemic playground of redemptive meanings. The works of Paul Willis (1990) and John Fiske (1989) are usually taken as the major instances of this kind of tendency. In Fiske's *Reading the Popular*, for example, he optimistically interprets a bumper sticker – 'A woman's place is in the mall' – as a form of ironic resistance to women's subordination to domesticity in the home. The shopping mall is a place 'where women can be public, empowered and free, and can occupy roles other than those derived by the nuclear family' (Fiske 1989: 18–20). Shopping itself is interpreted as a practice with subversive qualities; it values women's consuming skills and knowledge; the woman's spending of her husband's money [sic] is an act of conjugal resistance; shopping even has carnivalesque aspects.

Such an account has considerable problems. As Slater puts it:

> the undeniable fact that people are not 'cultural dopes' is taken to warrant investigating their active 'symbolic labour' as if it were unconstrained by social relations and necessarily subversive of them. Consumption always seems to happen in the 'gaps or spaces' – a realm of free self-determination – rather than in the guilty practices of mundanity; and these gaps or spaces always seem to be spaces of rebellion. The result is a continuous production of redemptive readings in which texts and objects are always viewed in terms of the spaces they offer for pleasures and fantasies through which people can 'make sense', and the consumer's redemptive readings will always neutralise whatever is 'bad' in the text.
>
> (Slater 1997: 169)

Our account is rather different from these two extremes. Taking its cue from the work of writers such as Mary Douglas, Pierre Bourdieu and Daniel Miller, it interprets shopping malls as a part of the process by which goods communicate, and are communicated as, *social relationships*. In turn, we see malls as places which perennially reconstitute these relationships through various practices of shopping and identity. In particular we follow Meaghan Morris's (1988) seminal article, 'Things to do with shopping centres', which decisively short-circuited the kind of thinking which frames consumption as either a seductively fallen state or a semiotic democracy. It did so in five main ways, to which we also subscribe.

First, her article refused to subscribe to the idea that all shopping malls were, at root, identical. For Morris, shopping malls have histories; she is aware that, 'if the shopping mall appears new and placeless today, this is because it has not yet been integrated back into its surrounding urban fabric, either by wear and tear, or by feats of the imagination, or by reputation' (Savage and Warde 1993: 143). Thus, whereas some of Morris's malls are new and gleaming, others are growing old and dowdy.[9] Like Brent Cross, their concrete is stained and they look 'seventies'. It follows that it is no surprise to find that, for Morris, each mall has a distinctive 'sense of place' and 'it isn't necessarily or always the objects consumed that count in the act of consumption but rather that unique sense of place' (Morris 1988: 194). As she says:

> Obviously, shopping centres produce a sense of place for economic 'come-*hither*' reasons, and sometimes because the architects and planners involved may be committed, these days, to an aesthetics or even a politics of the local. But we cannot derive commentary on their function, people's responses to them, or their own cultural production of 'place' in and around them, from this economic rationale. Besides, shopping-centre identities aren't fixed, constant or permanent. Shopping centres do get facilities, and change their image – increasingly so as the great classic structures in any region begin to age, fade, and date.
>
> (Morris 1988: 195)

Further, as Morris points out, this sense of place is increasingly played to. As more and more shopping malls appear so they are having to differentiate themselves from the others: a display of difference may well be an increasingly important part of the marketing strategy of shopping malls.

Second, performances by people in shopping malls constantly exceed the wishes or plans of their designers and managers.

> The stirring tension between the massive stability of the structure, and the continually shifting, ceaseless spectacle within and around the 'centre', is one of the things that people who like shopping centres really love about shopping centres. At the same time, shopping centre management methods (and contacts) are very much directed towards organising and unifying – at the level of administrative control, if not of achieved aesthetic effect – as much of this spectacle as possible by regulating tenant mix, signing and identifying styles, common space

decor, festivals, and so on. This does not mean, however, that they succeed in 'managing' either the total spectacle (which includes what people do with what they provide) or the responses it provokes (and may include).

(Morris 1988: 195–6)

They also often exceed the works of academics and other researchers since

you can't treat a public at a cultural event as directly expressive of social groups and classes, or their supposed sensibility. Publics aren't stable homogenous entities – and polemical claims assuming that they are tell us little beyond the display of political position and identification being made by the speaker.

(Morris 1988: 204)

Third, shopping malls cannot be seen as the habitat of the 'cruising grammarian', a *flâneur*-like shopper taking everything in and critically evaluating it with a sceptical eye. They can and do inspire complex and localised affective relations which can never conform to this stereotype. Sometimes, indeed, malls may be little more than 'managerial props for the performance of inventive scenarios' (Morris 1988: 223). More to the point perhaps, from Morris's feminist perspective such a stereotype entirely misses the fact that shopping is often hard and tedious graft: 'the slow, evaluative, appreciatively critical relation is not enjoyed to the same extent by women who hate the car park, grab the goods and head on out as fast as possible' (Morris 1988: 203).

Fourth, shopping malls have to be considered in relation to other sites of shopping. One of the ways they obtain their qualities is by contrast to the other sites at which shopping takes place: to an extent, the attraction of the malls relies on the existence of other, less salubrious sites. In turn, the perceived middle-classness of many malls has produced the development of new retail forms like the neo-arcade and the revamped city centre department store aimed at up-market shoppers.[10]

Then, fifth and finally, Morris promotes a catholic methodological strategy based on the principle of ambivalence towards the material that confronts her and her own position. She recognises the value of what she calls 'professionally-based informatics' – procedures for sampling shoppers, searching for exemplary figures, targeting user groups, and so on. But her own method is much closer to that of ethnography, and mainly consists of 'chit-chat' 'with women I meet in and around and because of these centres that I know personally' (Morris 1988: 208).

These five principles inform the work that we carry out in this book on shopping in Britain. Thus, first of all, our concern is with shopping malls as very different kinds of places, serving very different kinds of publics, a fact of which management is very aware, and makes constant allowance for. In Britain, the variation in the identity of shopping malls that comes from age and location is increasingly played to; as the number and size of shopping malls has increased, and with it competition, so the earlier malls are both forced to keep up with *and* differentiate themselves from the newer malls, through rebuilding and refurbishment, aesthetic strategies like new forms of signage and marketing campaigns. For example, Brent Cross, which we consider in detail in this book, has been facing severe competition as newer and larger malls open near it. Thus whereas in 1992 Brent Cross was in second place in DTZ Debenham Thorpe's annual index of retail trading; it was sixth in 1994 and ninth in 1996. The mall has responded by building an extension, by letting in more 'natural' light, and through marketing campaigns (see chapter 3). But, ironically perhaps, much of this response simply mimics current industry wisdom from the United States and so does much less to differentiate the mall than might be supposed:

> The recent spate of upscale-mall renovations rely liberally on the addition of skylights, greenery and fountains. In fact, I'll argue that such face-lifts have been undertaken precisely to attract the baby boomers – now the chief commanders of disposable income – who object to the generic placeless aura of the malls they grew up in. Build a fountain, they will come; developers install nature like a sign to affluent middle-class shoppers, saying this place is a real place, and it's for you.
>
> (Price 1995: 188)

Second, we take it as axiomatic that the ambience of shopping malls cannot be reduced to just the play of dominant social categories. Even in Britain, there is, to use one of Benedict Anderson's (1983) neologisms, no obvious 'unisonality' about shopping malls. The increasing diversity of British civil society has ensured that even shopping malls which may seem predominantly middle class, such as one of those we consider, are criss-crossed by all kinds of other divides which mean that they do not add up to one public. Third, we are sure that British shoppers, like Morris's Australian shoppers, cannot be framed as 'cruising grammarians'. As our evidence shows, for many of our shoppers for much of the time, shopping is as much an obligation and a chore as a pleasure, an activity which is periodically enjoyed but from which enjoyment is not expected as such. If critically evaluative skills are brought into play, it is relatively rarely for the purpose of play. Fourth, in Britain shopping malls are one of a number of different kinds of shopping that is undertaken. Whereas in the United States or Australia, malls are an integral part of the landscape, in Britain, where the density of malls is much less, they are still often only used periodically, in conjunction with many other forms of shopping. Fifth, and finally, our methodological strategy has been, like Morris's, a catholic one which mixes numerous methods, such as standard survey techniques, focus groups, and ethnography in an attempt to 'triangulate' the practices of shopping. Before outlining our methods, however, we provide a brief discussion of the history and development of the two centres.

2

HISTORY AND DEVELOPMENT OF BRENT CROSS AND WOOD GREEN SHOPPING CENTRES

Introduction

Brent Cross Shopping Centre and Wood Green Shopping City are both a product of the expansion and large scale investment that occurred in London during the 1970s. Prior to the building of Brent Cross Shopping Centre, there were no major shopping malls in London. Wood Green Shopping City was built to replace the dilapidated High Street shops in Wood Green. These centres epitomised the modernist philosophy of planned urban development: local and regional shopping needs were to be met by building self-contained complexes rather than by the gradual growth and development of urban areas that had previously occurred. While Brent Cross was a privately financed commercial development, the development of Wood Green required significant public involvement from the Greater London Council and the local authority, the London Borough of Haringey. This chapter begins by documenting the history and development of the two centres, highlighting their planning and development. This is followed by a description of the locality of the two shopping centres and a discussion of questions of local identity, urban regeneration and crime in Brent Cross and Wood Green.

The history and development of Brent Cross Shopping Centre

Planning for a major shopping centre in London began in 1959 by the Hammerson Group Investment Trust Ltd. It took until 1964 for a planning application to be presented to Barnet Council as the Hammerson Group had spent a considerable time finding the right site for development. Hammerson set up a subsidiary specifically for planning the development of a shopping centre in London as the company foresaw that there was considerable commercial potential to be realised from the government's 'overspill' policy of relocating people from inner London to the outer areas because of the poor quality of housing in the inner city. In addition, the 1950s and 1960s saw a rise in real income which was reflected in a tremendous growth in retail sales. The growth

31

in the sale of consumer-related and allied goods was far higher than that of food and household goods (Retail Planning Associates 1978).

Brent Cross was the only site which appeared to offer the essential requirements of availability of land and road access for large numbers of relatively affluent shoppers. The initial application for planning permission for a shopping and recreational centre in 1963 was withdrawn because the Minster of Transport considered that it would prejudice the free flow of traffic on the trunk roads in the area. The Greater London Council (GLC) had indicated its support for the development by writing it into their 1969 structure plan. The GLC had also identified the area of North West London, broadly coincident with the Borough of Barnet, as having very poor access to shopping in its retail strategy for North London (see Figure 2.1).

The inquiry relating to the building of Brent Cross reported that the impact on established shopping centres would only be to slow down an increase in retail trade. The trade for Brent Cross was intended to come from the North Western suburbs and the outer-metropolitan areas and thereby reduce pressure on the West End (Planning Application for Brent Cross Shopping Centre, Greater London Council 1968). The borough of Barnet strongly supported Hammerson in its plan, proposing 22 additional sites in the borough. The shopping centre was initially planned to open in 1973, but it was not until 1976 that the centre was finally opened to the public. The 1963 planning application included plans for a boating lake, crèche, service station and bowling centre but all these plans were subsequently shelved, the centre becoming a purely retail venture.

Brent Cross was unusual at the time in that it was the only shopping centre to have been built on a previously undeveloped site outside of an established shopping area and not being part of a new town. When it opened in 1976 Brent Cross Shopping Centre had 82 tenants with 800,000 square feet of retail space (74,000 square metres) on a site of 52 acres (2.15 hectares). There were 3,500 free car parking spaces (later expanded to 5,500 in 1985) and over 4,000 employees. Besides the two anchor stores (John Lewis and Fenwick), the centre included a Waitrose supermarket and branches of Boots, C&A, Marks & Spencer and W H Smith (see Plate 2.1).

Barnet Council let the site to the developers on a 125-year building lease. It was previously used for allotments and the rest was waste land including a disused greyhound stadium. Although the local allotment society had initially protested and there had been an objection by the owner of Brent Cross Garage, by the time of the inquiry these objections had been dropped. The record of the inquiry indicates a remarkably low degree of objections from the local community (Planning Application for Brent Cross Shopping Centre, GLC 1968).

Barnet Council benefited considerably from the building of the shopping centre as it had sold the leasehold to the Hammerson company but still received income from ground rent and a share of the rental income on the shops. In

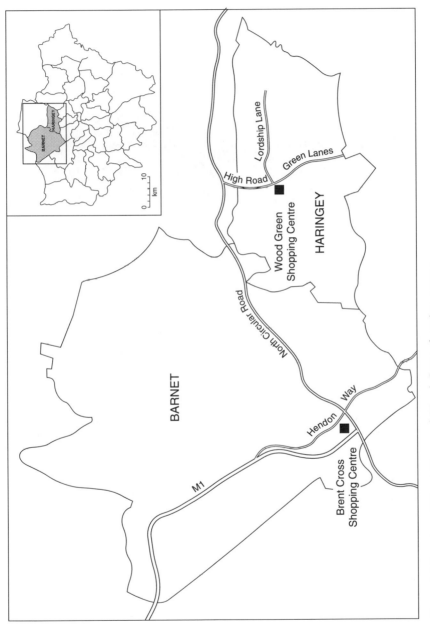

Figure 2.1 Location of Brent Cross and Wood Green shopping centres

Plate 2.1 Brent Cross shopping centre

1976, Hammerson was expecting an annual rent roll of £2–£2.5 million. The Council collected £650,000 in rates from the centre and £30,000 in ground rent, rising in five years to 16 per cent of rental income ('Brent Cross: £20m Gamble', *The Economist* 6 March 1976). Rents on the major shop units were individually negotiated and were related to the expected turnover of the shops. Fenwick's was said to have invested £4 million in its shop at the centre, being at the time the second biggest Fenwick's in the country. It also attracted new foreign investment. Other investors in the centre received a percentage of the rental income. For example, the insurance company Standard Life invested £20 million in the centre with a return initially agreed at 40 per cent of income for providing a 7 per cent mortgage. But there was considerable disagreement over this figure and £3 million of interest had to be paid off as quickly as possible. The development company's shareholders, Hammerson Property and Investment Trust with 75 per cent and Brent Walker (the leisure development company run by George Walker, brother of the former boxer, Billy Walker) with 25 per cent, were not expecting to profit on their investment for some years.

Many adjustments were made within Barnet to accommodate the new shopping centre. The Brent Park Road underground station was renamed Brent Cross station. About thirty bus routes were rerouted and the bus station was relocated to the centre. Local traders objected to these changes as they threatened to cut down their local custom. A year after the opening of Brent Cross further building was required. The initial plans for Brent Cross had not anticipated that a large proportion of its employees (over 4,000) would travel to work by car. Only 3,500 car parking spaces had been created, thought to be sufficient for customers and staff. The employees of Brent Cross were expected to park their cars in surrounding streets, causing considerable congestion and inconvenience having been deterred from using the customer car park by its later opening time ('The Revolution at Brent Cross', *Business Observer* 29 February 1976). There were concerns at the time that Brent Cross would be a 'white elephant' generating insufficient consumer demand. However, it proved to be very successful in its early years and this was reflected in the rapid expansion of car parking spaces at the centre. Despite what Gardner and Sheppard (1989: 102–3) call its 'design mediocrity' (a monolithic, block-like exterior with no natural light), the combination of free car parking and a retail-starved location ensured that Brent Cross became one of the country's most successful shopping centres.

Although there was no indication of serious disapproval by local traders and businesses prior to the development, there was considerable disagreement and objection once it was built. Traders in Hendon felt so strongly about the effect of the centre on local trade that the whole of the Hendon branch resigned from the Barnet Chamber of Commerce ('Traders Quit Joint Chamber', *Hendon and Finchley Times* 30 January 1976). There was also a vote of no confidence in the chairman of the Barnet Chamber of Commerce over this issue. The Hendon Chamber of Commerce accused the *Hendon and Finchley Times* of colouring editorial matter to encourage greater advertising revenue from Brent Cross and of being in the pocket of Hammersons, though it later had to apologise publicly for making this accusation against the newspaper ('Traders Sorry for the Slur', *Hendon and Finchley Times* 13 February 1976). In contrast, the Golders Green Chambers of Commerce were in favour of the development, seeing it as a way of improving the level of retailing in their area. Finchley traders reacted to the development of Brent Cross by organising a 'shop locally' campaign, distributing posters to shops in Finchley and Whetstone.

Brent Cross was planned on the assumption that it would draw shoppers away from the congested West End and not affect local trade. This argument was made on the basis that there would be a concentration at the centre on consumer durables and comparison goods rather than food and household goods. Hendon traders did not believe this and saw Brent Cross as a major threat to their livelihood. At the time the traders were regarded as unreasonable, but subsequent developments seem to have substantiated their early suspicions. The Brent Cross shopping centre manager agreed that the centre had attracted quite a lot of local trade from Hendon which had seriously affected the vitality of the town centre (interview, November 1993). Most recently, a sharp decline in the fortunes of Golders Green retailers, symbolised by the spread of charity shops, has also been blamed on competition from Brent Cross.

A report commissioned by the shopping centre in 1988 indicated that the number of people visiting Brent Cross Shopping Centre was declining. The data also indicated an ageing of the customer profile since 1979 and a reduction in 18–25-year-olds. The centre was becoming outdated. The 1970s design and architecture had become unfashionable and other shopping centres were attracting trade away from Brent Cross. Lakeside, on the M25 motorway in Thurrock, for example, has over 1 million square feet (93,000 square metres) of shopping space compared to Brent Cross's 800,000 square feet (74,000 square metres). Brent Cross was also facing competition from the Galleria in Hatfield which, although smaller than Brent Cross, catered for upmarket shoppers. The proposed Blue Water development in Dartford was also regarded as a major threat. Lakeside had thrown down the gauntlet by advertising on a poster site directly opposite Brent Cross ('Complex Issues: Brent Cross is Facing Serious Competition from around the M25', *Hendon and Finchley Times* 27 February 1992).

In comparison with its later competitors it had five main problems. First, it was too small. Second, the flow of pedestrians was restricted by two escalators and a single lift. Third, there were few places to sit and eat with only two cafés and two fast-food outlets. The result was that average stay times in the mall were only of the order of two to three hours. Fourth, there was a lack of car parking. Fifth, the decor felt dated. The architect in charge of the subsequent refurbishment argued that it conformed to the 'disco aesthetic' of the 1970s with artificial lights, marble, and so on (based on the idea that if the central areas were dark the customers would be attracted into the brightly lit stores).

Responding to these threats, Hammerson announced at the end of 1991 that they were applying for permission to extend the centre ('Shopping Around for Extra Room', *Hendon and Finchley Times* 12 December 1991). A larger plan had been submitted to Barnet Council in 1986 but this had been rejected because it was felt that the local roads and car parks needed to be improved before any extension was agreed. The second proposed extension was smaller to make more room for additional car parking. Plans were submitted in January 1992 and the refurbishment began in October 1994. As well as adding ten extra shops, Fenwick's and Marks & Spencer were applying to extend their shops ('Plans for a Bigger Brent Cross Shopping Centre', *Hendon and Finchley Times* 9 January 1992). In the event the overall size of the mall was expanded by including a 50,000 square feet extension, costing around £80 million (see Plate 2.2), which also allowed the overall retail mix to be changed, with a bias towards more upmarket clothes shops in particular. Second, the pedestrian flow was increased by providing eight escalators and two glass lifts. Third, the number of cafés and restaurants was increased by adding a food area into the upper level of the new extension with five individual operators. Fourth, car parking was increased through the construction of a new multistorey car park (and the demolition of the old one) adding 2,000 new parking spaces. Fifth, the decor of the mall was changed. Specifically, the low ceiling over the central mall was replaced with glass in order to let in more natural light – even the showpiece plaster dome over the central court was replaced with glass in pursuit of this aim. All of these changes were completed by 1996.

Plate 2.2 The refurbishment of Brent Cross Shopping Centre

The history and development of Wood Green Shopping City

Wood Green is one of the larger shopping centres in London and contains the administrative centre of the London Borough of Haringey. As with Brent Cross Shopping Centre, plans were under way to build a centre at Wood Green in the 1960s. Earlier this century, Wood Green had a very strong co-operative movement with the largest co-operative society shop in London located in Wood Green in 1947. There were many smaller co-op shops in the area before this. The area was reasonably up-market, regarded locally as the 'Golden Mile of North London'. There were several department stores in the area such as Barton's and Edmund's. However, with the closure of the Alexandra Palace Gate railway in the 1960s much of the centre of the town had become rather down at heel and many shops and other buildings in the centre were demolished during this period (Woods 1991).

Before the building of the Shopping City, Wood Green was characterised by a number of Victorian and Edwardian Parades many of which disappeared with the construction of the centre. The main shopping parade was the High Street at one end of which Wood Green Shopping City is now located. The library was constructed next to the Shopping City at the end of the old parade, with the High Street now dominated by the bridge which crosses over the road, linking either side of the centre (see Plate 2.3).

Unlike the free-market ethos which dominated the development of Brent Cross, the building of Wood Green Shopping City was strongly influenced by a sense of municipal co-operation between the council and private investors.

Plate 2.3 Wood Green Shopping City from the high street

The council believed that it could combine its role in providing council housing and leisure facilities with a scheme that would modernise the town centre and provide a new municipal focus for Haringey. The reorganisation of local government in London in 1965 provided Haringey Council with the opportunity to develop Wood Green as 'the coherent town centre in the heart of a new borough' (Wood Green Planning and Development Committee 1965). Previously, there had been many ambitious plans for the redevelopment of Wood Green such as the building of an arcade over the High Street, with the public conveyed to pedestrian areas via an electric mini-rail system linked to the bus and underground stations. Such plans were shelved with the creation of Haringey Borough from an amalgamation of Hornsey, Tottenham and Wood Green councils and new plans began to emerge for the building of a library, recreation centre and theatre. The High Road itself was to be pedestrianised and landscaped.

A major exhibition was held in 1973 to explain the schemes to residents. It was visited by the Queen Mother and included a model of the town centre. Wood Green Shopping City was intended to comprise 500,000 square feet (46,000 square metres) of shopping space and 1,566 car parking spaces. The centre was designated as one of London's six regional shopping centres by the GLC. Plans for a leisure centre and theatre were subsequently dropped. A public house was also planned, as well as a crèche for shoppers and residents of the housing scheme which was built on top of the Shopping City. Attempts were made to blend in the buildings with the locality by using a Southwater brick throughout, and an overhanging first floor on all external frontages was intended to give weather protection, add to the sense of integration of the scheme and reinforce the relationship between the adjoining pavements and the High Road.

Electricity Supply Nominees (ESN) were chosen by the council to run the scheme from 50 firms who bid for the contract, leasing the Shopping City site and providing the council with a share of the rental income from the shops. ESN also pay a minimal ground rent to the council. The housing deck above the centre is leased back to the council which in turn underleases this property to the Metropolitan Housing Trust. A public inquiry was held regarding the Shopping City in 1972 and Phase I of building was finished in 1977. A temporary market was constructed in this building for displaced traders from the original town centre. The second phase was built between 1976 and 1979. The bridge and the market in Phase II was opened in 1979 and the malls and the majority of the shops were completed in 1981. The housing developments were also due for completion at this time. Housing above the shopping centre, in three- and four-storey terraces, provides 201 units, ranging from five-person maisonettes to one-person flats. A lift provides a direct access to the Shopping City. It has since been found that many of these units have been hard to let and there have been many environmental problems associated with living above the Shopping City (interview with Wood Green Shopping City manager, November 1993).

Wood Green Shopping City was marketed by the Richard Ellis Company as providing the 'Golden Heart of Retailing in North London' ('Wood Green Shopping City: a development by Richard Ellis', marketing brochure, date unknown). The logo for Wood Green at this time was a golden heart with a sparkling reflection. The publicity brochure makes reference to Wood Green's history, going back as far as the middle ages and reproducing sepia photographs of Wood Green at the end of the last century. The slogan at the end of this section of the brochure states that 'Wood Green Shopping City profits from its position because it profits from its past'. It was opened in 1982, with an earlier visit by the Queen who unveiled a plaque in the market square (*Weekly Herald* 7 and 14 May 1981). The housing complex was described by the *Herald* as 'The Village in the Sky' and was claimed to be a new concept in housing. Every resident would have a 'kind of cottage' and youngsters could play in the village squares. The philosophy of the developers, according to the *Herald*, was to try to make shopping fun rather than a chore, promoting peripheral activities which aimed to attract and amuse shoppers. Their slogan at that time was to 'Put the heart back into shopping', implying that the Shopping City would renew the economic and social vitality of the area.

Although the GLC designated the proposed Shopping City as a regional shopping centre, since its opening it has attracted a very local customer base. In hindsight, the shopping centre may have been too large for its catchment area, especially as more and more people began to travel outside of the immediate area to shop. Unlike the relatively self-contained Brent Cross, Wood Green Shopping City has also had to compete with shops on the High Street.

The shopping centre lacks the anchor stores that Brent Cross has, with only one major department store, Evans, facing direct competition from Marks & Spencer across the High Street. But the most serious problem with the planning of Wood Green town centre has been the inability of successive local governments to tackle the traffic congestion in the town centre. A bypass was planned and later abandoned for lack of money. Meanwhile, the car parks have become local crime spots and are regarded by many local residents as quite dangerous places to park.

The errors in the planning and design of the centre were so serious that the shopping centre surveyors, Richard Ellis, were sued for wrongly assuming that the Shopping City would draw people from all over London. In fact, the centre has come to rely on a very local clientele drawn from mainly Wood Green and Tottenham. One informant told us that Richard Ellis had been responsible for all of ESN's investment portfolio and had also been sued by them for serious errors in the design of the Trocadero in Leicester Square. The case was settled out of court with ESN receiving substantial compensation.

After only eight years trading, a refurbishment of Wood Green Shopping City was undertaken in 1989 at a cost of £5 million. The number of shops was reduced through combining some of the smaller units into larger stores. The shop frontages were all changed and new kiosks were built downstairs. The floor was also replaced and the bridge across the High Street was renovated. According to Kit Horntvedt Associates, who were brought in to create a new image for Wood Green Shopping City, the centre was 'too dark, too institutional and too unfriendly for modern tastes' ('Shopping Centres Need a Brand Image', *Shopping Centre Horizons* May 1990: 12). Market research had indicated that only shoppers under 25 were prepared to spend time browsing and shopping at the centre. The car parks had become a serious problem and people were not prepared to leave their cars in the vicinity. The refurbishment was planned to move the Shopping City more up-market while being careful not to compete with Oxford Street or the likes of Brent Cross.

The old red and green logo was described as 'reminiscent of the Waffen SS' and was redesigned to coincide with the renovation. The new rainbow tree logo, with its implication of healthy multiculturalism, was intended to 'capture the excitement, fun and style that people looked for . . . when they went to Shopping City' (ibid.: 12). Twelve new shops were also incorporated within the centre ('5 Million Facelift for Shopping City', *Haringey Advertiser*, date unknown). At the same time, a new manager was appointed with the task of maintaining the Shopping City as a separate entity from the High Street. (In practice this has been hard to achieve as many of our informants found it difficult to distinguish between the High Street and Shopping City.) The redesign of the town centre has contributed to major traffic congestion and parking problems which many people mention in regard to Wood Green.

Despite the planning and design problems, Wood Green Shopping City has quite a popular image locally. Many events have been organised to promote the Shopping City within the locality such as money-back schemes to encourage people to shop there, fashion shows and multicultural activities. The centre markets itself as being associated with a kind of popular, middle-of-the-road image as represented by television and sporting celebrities such as Barry McGuigan, Suzanne Dando, Michael Fish and Ruth Madoc who visited Wood Green in 1994 or members of the Saturday night TV show *Gladiators* who made personal appearances there in 1995. Notwithstanding its popularity with local people, since the late 1980s Wood Green Shopping City has seen a number of shops close and there is little sign of the centre attracting new shops and businesses. Recently the Shopping City took over control of all the centre's car parks from Haringey council in an attempt to improve parking facilities throughout Wood Green. Advertising for the centre now focuses on locality, convenience and difference (see Plate 2.4). Most recently Wood Green Shopping City has been sold by ESN to a new owner, Capital and Regional Properties.

Plate 2.4 Local advertisement for Wood Green Shopping City

The mall business

As they have become more common, so shopping malls have become a distinctive segment of the retail industry. Though they are now much more numerous, shopping malls are still a relatively recent development in Britain and many areas of the country are still outside the range of a major mall.[1] Though it has been clear that running a shopping mall is not like running, say, a department store, until recently malls have often been subsumed into the more general structure of retail business. However, as we shall see, a distinctive circuit of mall ownership and management is now beginning to emerge, with its own knowledges and skills.

This circuit is based upon three main sets of actors. The first are the owners, usually property companies or institutional investors. The second set of actors are the managers, who are usually acting as agents of the owners. These tend to be the familiar large estate agents such as Richard Ellis, DTZ Debenham Thorpe, Donaldsons, and so on. Then, finally, there are the retailers leasing space in the malls. We now consider these three actors in turn.

The mall owners

Shopping centres such as Wood Green and Brent Cross have proved highly attractive to investors. For example, they made up 16.8 per cent of institutional property portfolios in 1995, up from 9.4 per cent in 1981. Their attraction is based on above-average retail rents, and the fact that their very size means that they are a coherent investment package. It is no surprise, then, that nearly 200 shopping centres changed hands between 1990 and 1995.

Most of the shopping centre owners are property companies and large institutional investors. For example, Wood Green has been owned since its inception by ESN (Electricity Supply Nominees) Pensions Management, the fund manager which helped to develop it. But in 1996, it was bought for £33 million by Capital and Regional Properties, as part of this company's plan to increase its exposure to the retail sector (since 1994, it has also bought centres in Aberdeen, Newcastle and South London). Brent Cross, by contrast, was developed and has been owned since its inception by Hammerson, the third largest property company in the UK, which also has extensive overseas interests (with 60 per cent of its assets in Australia, Canada and the United States) and, in effect, Standard Life, the life assurance company, which is a large shareholder in Hammerson.[2]

In recent years, the real challenge for investors has been the lack of good-quality shopping centres that do not require intensive management or refurbishment, a problem exacerbated by the increasing government bias against out-of-town developments (discussed in more detail in chapter 4). The result has been a flurry of refurbishment activity as one of the only ways to add value to retail portfolios. Thus the main mall areas of Wood Green were refurbished in 1989–90, at a cost of £5 million, and as we shall see below, Brent Cross was refurbished between 1994 and 1996 for considerably more.

Most malls are managed by agents on behalf of the owner, normally London-based, who specialise in shopping centre management for a fee. Wood Green is managed by DTZ Debenham Thorpe while Brent Cross is managed by Donaldsons. At Wood Green, the manager of the mall is positioned between the managing agents and the owners. In Brent Cross, the manager is a direct representative of the managing agents (who, if he was located at head office, would have full-salaried partner status) and reports directly to an equity partner.

42

The mall managers

The malls are run by small management teams which are employed by the mall's agent. In turn, these teams are responsible for a number of management tasks. Chief amongst these have been managing the service charge levied on tenants (which is mainly used to make sure that the mall is kept tidy, secure and in good condition) and liaising with tenants. Both of these are demanding tasks. The malls have to employ sizeable work forces to ensure that the mall is tidy and secure. For example, in 1995, as well as the management team, Wood Green employed 19 cleaners, 16 security staff, and a full-time handyman. The manager of Brent Cross was responsible for 120 staff. Security is a particular concern and the malls both have a large security force and closed-circuit television systems centred on a control room.[3] But it is liaising with tenants that managers find a particularly contentious task. Tenants routinely complain that the service charge is too high. If tenants lose revenue, they tend to blame the mall management, a problem which is compounded because local retail managers often have little freedom of action and often take their instructions from higher up the chain of command. Then again, some shops may become shabby and run-down, in which case managers must take diplomatic action to ensure standards. Mall-wide improvements like new signage often have to be agreed with tenants. Then, if promotions are required, this may well involve negotiating with tenants for extra resources since, until recently, the mall's promotional budgets have been relatively small. The formula which has often been applied by management in British malls of 'high rents, low charge' is directly responsible for many of these woes, since it encourages tenant isolationism.

But promotion and marketing has now become a third major management task. Both centres have began to market themselves extensively, not only in the press but also on television (for example, Brent Cross used the TV personality Ruby Wax in a pre-Christmas campaign). Both centres now routinely run promotional events, for example fashion shows, or appearances by celebrities. Wood Green has used comic actors such as Ruth Madoc and Paul Shane, although their greatest success was the boxer Nigel Benn, who brought in a very large crowd. Brent Cross has used the TV presenter Anthea Turner to switch on the Christmas lights. As a result of marketing efforts like these, the malls have become much more conscious of their image and strive for a particular kind of ambience. For example, Brent Cross strives for an up-market image. It uses popular classical music as background, its security staff are dressed like police officers and are clearly visible, and so on. By contrast, Wood Green background music tends to be more eclectic, security staff are dressed in maroon jackets and black trousers so that they do *not* give the impression of being associated with the police, and the Centre has developed a logo which stresses its multicultural ethos.

This increased marketing effort is, in part, associated with a change in the nature of mall management. Until recently mall managers tended to come to the job from diverse backgrounds (but with a bias to particular backgrounds, such as the military). Now, with the growth of shopping centres and malls, there are sufficient opportunities to have produced a new branch of the retail division of labour, the shopping centre manager, with a relatively distinct set of skills and even a career structure.[4] Thus, within the managing agents, managers can move up a career hierarchy. For example, in 1995, Donaldsons, the Brent Cross agent, was managing 51 shopping centres spread across the country. Managers also meet regularly at an annual conference in Oxford and there is now even a diploma in Shopping Centre Management (for which one of the deputy managers at Brent Cross was studying).

The increased emphasis on marketing, as with much else in the world of mall management, has North American roots. This is no particular surprise.[5] After all, the history of British retailing is littered with American imports, not the least being the example of the opening of Selfridge's in 1909, an operation run by Gordon Selfridge on the lines of Marshall Field's in Chicago which he had so precipitately left, and based on retailing skills of an American controller of merchandise, an American organiser of layout and finishing, and an American designer and window artist, as well as Selfridge's own showmanship (see Fraser 1981; Lancaster 1995). Brent Cross is a particularly good example of how retailing ideas constantly cross the Atlantic and become embedded in a new milieu. Brent Cross was based on a standard US mall design of a 'dumb-bell' shape, with two anchor department stores – John Lewis and Fenwick's – linked by a two-level central mall.[6] When the mall opened in 1976, it was state-of-the-art but less than 20 years later it was already feeling its age. Almost all the changes that were made in the 1996 refurbishment, such as letting in natural light and the glass lifts, are in line with standard contemporary transatlantic retail thinking on mall design and decor.

The North American influence also extends into other areas of Brent Cross's life, particularly marketing. And there are regular contacts between Hammerson's North American and British retail operation that ensure this. For example, selected staff are sent across 'territories' to shadow foreign colleagues:

> The first [in Britain] to be chosen for this mission was Sheila Sullivan, Hamerson's retail development director in the UK. A surveyor who joined two years ago after a stint at Capital and Counties, Sullivan was teamed up with the vice-president of Hammerson's Canadian retail operations, Howard Quennell.
>
> Sullivan didn't literally dog Quennell's footsteps but she did haunt Square One, Hamerson's great centre in Mississauga outside Toronto, Canada, and the corporate head office opposite . . .
>
> In fact, Sullivan's stint kicked off in Las Vegas, attending a large retail leasing fair with Quennell. This allowed her to meet some of the new American retailers that she wanted to lure to London's Brent Cross. 'We did meet quite a few and managed to do a few deals on

the back of that, so that was a good start to the shadowing process',
she says. As a result US retailers Talbot and the Museum Co are
moving into Brent Cross.

(Catalano 1996: 50)

The general emphasis on marketing and customer service in North American
malls is having a particular impact on Brent Cross. Thus, the mall now has a
marketing manager, part of whose brief is to bring what are largely North
American ideas into Brent Cross.

The mall retailers

In malls there is a constant struggle between managers and retailers. In this
struggle retailers have considerable negative power. They tend to be on long
leases and fixed rents and these make it difficult for managers to actively manage
the retail mix in a mall:[7]

> The long lease maybe works to your advantage in a declining centre
> but not in a strong one where sales growth is occurring. American
> style turnover rents and shorter leases provide scope for more active
> management.
>
> (Catalano 1996: 51)

The manager can also find it difficult to draw up extra support for new intia-
tives which involve contributions from numbers of tenants. On the other hand,
tenants do not have much positive power. The exceptions, of course, are the
anchor department stores and the growing number of larger retail chains like
C&A, W H Smith and Boots (which are present at both our malls) who clearly
do have more clout. These stores and chains can make life difficult for managers
since they can go round them to head office.

What is clear is that, until recently, the relationship between the managers and
retailers was one at arm's length. This is in marked contrast to the United States:

> tenant initiatives are encouraged in North America. Indeed the rela-
> tionship between landlord and retail tenant is much more developed
> across the pond. In Canada, Hammersons runs conferences and courses
> and has sessions with 'store doctors' for tenants. These mainly benefit
> the smaller 'mom and pop' independents.
>
> 'It's in our interests to spend a little bit of money to make these
> tenants better. And it's much cheaper to be able to review a tenant
> at the end of a lease than to say "sorry it hasn't worked" and go
> through a vacancy and then rebuild that space out for another tenant',
> says Quennell.
>
> (Catalano 1996: 51)

45

But the times are now changing. Tenants are more able to take the intiative and managers are more likely to want them to. The move to a modified North American model of the landlord-tenant model seems likely.

The management task is clearly very different between Wood Green and Brent Cross in certain aspects. The manager of Wood Green has had to deal with a vacancy rate as high as 10 per cent over the course of its history and has hardly, therefore, been in a position to pick and choose tenants. In contrast, Brent Cross has nearly always been over-subscribed, with many retailers wanting to get in. Now, of course, with the increased competition from new malls at places like Lakeside, this may well change.

The Brent Cross and Wood Green localities

Given its suburban, out-of-town location and its intended 'regional' focus, Brent Cross Shopping Centre has a surprisingly local catchment area. Data from our questionnaire survey indicate that quite a high proportion of Brent Cross shoppers come from the borough of Barnet, in which Brent Cross is situated. Although only 17 per cent of Brent Cross shoppers are from Hendon and Golders Green, the nearest suburban centres, this rises to over 40 per cent when the other postal areas in Barnet are included. Two other boroughs also have a reasonably high proportion of Brent Cross shoppers: Harrow with almost 10 per cent and Brent with 8 per cent. None the less, Brent Cross draws its clientele from a much wider area than does Wood Green Shopping City. A high percentage of Wood Green shoppers come from the immediate area, with a majority (55.2 per cent) from the Haringey borough.

Haringey, an inner-city borough, has for many years been a Labour strong-hold and has a significant concentration of people from ethnic minority backgrounds: 29 per cent compared to just over 20 per cent for Greater London as a whole (*Regional Trends* 1994). There is a particularly high concentration of Black Caribbeans in the borough (9.3 per cent) in comparison to Greater London as a whole (see Table 2.1). Haringey's cultural diversity is also illustrated by the number of languages spoken in its schools. An Education Service Survey of Haringey schoolchildren found that they spoke 193 languages (*Haringey in Figures* November 1993). Turkish (2,109 households) was the most widely spoken language, followed by Greek (1,112 households) and Bengali (991 households). The proportion of unemployed people in the borough is the fourth highest in Greater London (*Regional Trends* 1994) and there is a low percentage of economically active people in the area. Haringey recently applied to be an Objective 2 region within the European Union (EU) because of its economic problems although its application was not approved because the borough also includes the wealthy area of Highgate. In 1991 Haringey had the second highest level of homelessness in Greater London together with the fourth highest proportion of unfit local authority dwellings. However, property prices are just below the average for Greater London. The

Table 2.1 Ethnicity of residents in Haringey

Percentage	White	Black Caribbean	Black African	Black Other	Indian	Paki- stani	Bangla- deshi	Chinese
Greater London	79.8	4.4	2.4	1.2	5.2	1.3	1.3	0.8
Outer London	83.1	2.7	1.3	0.7	6.5	1.4	0.4	0.7
Inner London	74.4	7.1	4.4	2.0	3.0	1.2	2.8	1.1
Haringey	71.0	9.3	5.5	2.3	3.6	0.7	1.5	1.1
Barnet	81.6	1.0	2.1	0.5	7.3	0.8	0.4	1.3
Brent	55.2	10.2	4.1	2.2	17.2	3.0	0.3	1.1
Harrow	73.8	2.2	0.8	0.7	16.1	1.2	0.3	0.9

Source: OPCS 1991

average price of a property in Haringey in 1991 was £83,200 compared to £88,700 for Greater London (London Housing Statistics, London Research Centre 1992). Haringey therefore provides some interesting contrasts between the poor and the privileged, having many of the characteristics of a poor inner London Borough but also some areas with a relatively high proportion of middle-class residents.

The London Borough of Barnet, in which Brent Cross is located, was until recently Conservative-controlled but in 1995 was in the hands of a Labour and Liberal Democratic alliance. The borough is characterised by a strong predominance of White residents (over 80 per cent, although the Census does not identify what is probably the largest concentration of Jewish residents in Britain), the next highest ethnic category being of Indian origin (7.3 per cent). There is a relatively low percentage of Black African (2.1 per cent) or Black Caribbean (1.0 per cent) origin. Census data on social class for Barnet indicates that the percentage of people in Class 1 (employers and managers) and Class 2 (professional workers) is over 30 per cent compared to just under 25 per cent for Greater London as a whole. Unemployment rates for Barnet are consistently 4–5 per cent lower than for Greater London. Almost 70 per cent of residents are in owner-occupied dwellings, compared to 57 per cent for Greater London. Barnet also has low levels of homelessness and a very small proportion of council properties in need of renovation.

Residents in Haringey tend to work in other parts of London or in service industries in the local area. There is some light industry in Tottenham including engineering, clothing and confectionery making. Many commute to the West End or work in the City of London. It is a fairly low- to middle-class area, though with a growing proportion of upwardly mobile people who have moved to Wood Green to move up the property ladder. Barnet, on the other hand is much more solidly middle class and very residential in nature with many people employed in service sector industries such as banking and finance.

Local identity, urban regeneration and crime

When Brent Cross Shopping Centre opened in 1976, the *Evening Standard* referred to it as the 'Housewife's Dream', a place which exemplified a new form of shopping where a washing machine, a new dress or a pound of apples could be bought in the same place (6 January 1976). The centre was marketed towards the middle-class family and concentrated on meeting the needs and aspirations of the women within these households. Generally, however, the reception of Brent Cross by the public was more ambivalent. Commenting on Brent Cross two years after it opened, the following contradictory assessment was not uncommon:

> The massive and featureless rectangles of the Brent Cross Shopping Centre ... come as no surprise. They are just as hideous as every-thing else. But step through the doors and here is prettiness and femininity – just as soulless and just as commercialised as the filth outside but a veritable perfumed nirvana.
>
> ('Nirvana with Shops', *New Society* 2 February 1978)

Later in the article, the Centre is described as 'the woman's world of the second half of the 20th century', ameliorating the sense of social isolation and lack of facilities for mothers with children experienced elsewhere. A 'hothouse' metaphor is used to highlight the sense of the exotic at Brent Cross, where housewives will flourish 'as do the poinsettias, pineapple plants and rubber trees in an environment kept constantly at 60 degrees centigrade'. The article continues, 'As she pushes her pram up the main mall, the average shopper (25–34), moves as if being filmed for a building society advertisement. She talks to her offspring exactly as Dr. Spock says she should.' It is never ques-tioned that this is a women's world or that the predominant context is that of 'family shopping' (see chapter 5).

By 1989, Brent Cross was being described in the *Hendon and Finchley Times* as 'the place that we love to hate'('Shop around in the Nineties', 29 December 1989). It was termed a 'Mecca' for fair-weather shoppers, but could induce high blood pressure as shoppers battled for an empty car parking space. Similarly mixed feelings were expressed by the Chairman of the Hendon and District Chamber of Commerce:

> Brent Cross has affected the shopping trends in the area. The Centre was jointly opened by Barnet Council. The law changed months before it officially opened so instead of a five and a half day shopping week, it could trade over a six day week. In later years the centre was opened until 9pm. Nothing was ever changed to help the small shopkeepers. Offices, building societies and estate agents have sprung up everywhere changing High Streets from retail to non-retail all too easily and

absolutely ruining areas like West Hendon, Hendon Central and more recently Brent Street.

Recent plans to extend the centre have generated much more vigorous objection than was expressed towards the original plans. Local residents objected to the uprooting of trees for new car parks, arguing that they formed a sound and pollution barrier ('Residents Vow to Fight Plan to Expand Shopping Centre', *Hendon and Finchley Times* 28 April 1994). The Brent Cross Residents' Association (who participated in our focus group research) had very strong views about the centre's effect on the local area. The Association alleged that the Council was in the pocket of Hammerson before the recent elections when the Conservatives lost control of the Borough. Of particular concern was the level of pollution in the area, increasing levels of traffic congestion and litter, and shoppers parking in the area. They had previously felt powerless to negotiate with Hammerson and Barnet Council as they often did not know of proposed developments until they were about to happen. With the political changes in Barnet Council and the Unitary Development Plan requirement that the council take notice of local pressure groups, they found that Hammerson was now much more prepared to listen to their views. During the recent refurbishment they had met with the Directors every month to discuss the changes, but they still found that Hammerson was liable to change its plans without prior consultation.

Residents who lived near Brent Cross felt that the shopping centre had a very profound effect on the vicinity. In their view, Hendon Central had been virtually closed down by Brent Cross and the only shops left on the High Street were building societies or funeral parlours. They thought that even the High Street at Burnt Oak, on the edge of the borough, had been negatively affected by the building of Brent Cross. The Association were largely of the opinion that the shopping centre had little character and had done little to foster a sense of community in the local area. They complained about the artificial nature of the shopping centre, the poor lighting and air conditioning. They were, however, quite fond of the fountain and the stained-glass ceiling and were saddened that it would be removed with the refurbishment.

A far stronger sense of local identity existed in Wood Green than was evident from the focus group research and other evidence in the vicinity of Brent Cross. The sense of locality was illustrated by the number of common places, memories and incidents mentioned by people in the focus groups and the concern about the general upkeep of the area. Some of the older people who participated in our research had been relocated because of the building of the shopping centre and had experienced dramatic changes through their lifetime from the replacement of old parades and markets with shopping malls and department stores. Attitudes to the Shopping City were therefore related to perceived changes in the local social and cultural fabric. Associated with this was a profound sense of loss experienced by many of the older people. They talked about how much they had enjoyed parading along the High Street in Wood Green, going window shopping or chatting to the barrow boys who called out to them from the street corners. As we shall see in chapter 5, the loss of 'community spirit' was frequently related to the racialised nature of neighbourhood change due to the influx of people from ethnic minority backgrounds with which the construction of the Shopping City coincided.

Facilities for older people in Wood Green Shopping City are quite poor. During the refurbishment a lot of the seats in the shopping centre had been removed and older people were quite nervous about some of the modern amenities such as escalators and lifts. Some also found the layout of the shopping centre quite disorienting. Older people were also the most sensitive to the perceived level of crime and incivility in the neighbourhood. Many were fearful of being mugged in the streets or having their bags snatched at the bus stop.

Younger people in Wood Green had very different experiences, often having very positive associations with the Shopping City. For them, the centre had a particular vitality as a meeting place, somewhere they went to see and be seen. Wood Green is a particularly significant place for young Greek and Cypriot people because of the presence of the Greek Cypriot Centre there. Greek Cypriot youths saw the centre as their local meeting place, but there was a negative association with being seen at the shopping centre (as we shall argue in chapter 8).

Earlier studies of Wood Green have also concluded that the area has a poor sense of identity, regarded as a dark and threatening place in the evening. A report by the Urban Design Action Group (UDAG) and the London Borough of Haringey identified a number of problems in Wood Green (Wood Green UDAG 1990). Businesses complained about the poor upkeep and the lack of civic places in Wood Green where people could meet. They found that local people were particularly concerned about the heavy traffic in the vicinity and the problems of car parking. In 1993, a town centre manager, Richard Thomas, was appointed to try to improve the town centre environment. Located in Marks & Spencer, he was employed jointly by Haringey Council and by the shops on the High Street and in Wood Green Shopping City, with financial support from the Wood Green Town Centre Initiative. One of his main objectives was to improve the traffic conditions in Wood Green and to upgrade the multi-storey car parks. He also had responsibility for improving safety and security in the area, trying to get closed circuit television in the High Street.

A major initiative under way in 1995 was the plan to improve Spouter's Corner opposite Wood Green tube station which is a particularly unattractive area associated with a high rate of pedestrian accidents and traffic congestion. A new marketplace has also recently been developed in this area but it has not proved very popular with shoppers. Only 2 per cent of our survey respondents went to this market on a regular basis. Remarks were also made by teenagers about the 'nothing stalls' in the market and the lack of local identity to this place (cf. Jackson and Holbrook 1995).

Wood Green is known locally as a relatively dangerous place to live. This reputation is only partly born out by the available crime data, and crime levels actually appear to be falling locally. In 1991, data on violence against the person for all boroughs within London suggest that Haringey has one of the lowest incidences for this particular offence. In Tottenham, in particular, there had been a significant reduction in reported burglaries and street crime. Tottenham also has one of the highest clear-up rates for burglary in the whole of London (*Haringey People* November 1993). Both the town centre manager and the manager of Wood Green Shopping City felt that the local and national media have tended to sensationalise the level of crime in the town centre. However, there have been a number of violent incidents which have given Wood Green a bad name. In 1991, four police officers were stabbed by a man from a local mental hospital. Onlookers were reported to have jeered as the man stabbed three young police women and a male police sergeant ('Police Hurt in Stabbing "Jeered by Onlookers"', *Independent* 31 March 1992). There have also been a number of terrorist incidents in Wood Green including an explosion outside the Shopping City in December 1992. Bombs were hidden in rubbish bins outside W H Smith and Argos and ten people suffered minor injuries when they exploded ('Police and Shoppers Injured in Bomb Blasts', *Independent* 11 December 1992).

The level of crime in Barnet is generally much lower than in Haringey. The highest figures are for vehicle crime (with 6,745 offences in 1993). Brent Cross also experienced a major terrorist incident in 1991 when two bombs exploded during the centre's busiest period on the last Saturday before Christmas ('Fire Bomb Blitz', *Hendon and Finchley Times* 19 December 1991). The significance of these incidents, and of the perceived links between crime, race and neighbourhood change, are developed further in chapter 5.

Conclusion

The planning and development of Brent Cross and Wood Green Shopping City have been driven by very different political imperatives. Brent Cross Shopping Centre was first and foremost a commercial project initiated by the property developers, Hammerson who sought the co-operation of the borough of Barnet in building Brent Cross. As Britain's first planned out-of-town shopping centre, Brent Cross has gradually become a more local shopping centre, generating ambivalent feelings among those who shop there or live nearby. As our focus group and ethnographic work will show, different social and ethnic groups have developed varying degrees of attachment to the place.

In contrast, the development of Wood Green Shopping City was initiated by Haringey Council and planned to meet the social needs of local people as well as fulfilling various commercial objectives. Despite a sense of attachment to the centre's vitality and multicultural appeal, Wood Green has experienced a number of problems associated with traffic congestion, relative deprivation and high crime levels. Wood Green Shopping City attempts to be much more part of the local community than does Brent Cross but is hampered by a lack of resources and by the marked social differences within the neighbourhood (on whose significance we shall comment in later chapters).

3

METHODOLOGY

Introduction

As outlined in the Preface, this study aims to 'ground' the relationship between consumption and identity through an empirical study of two North London shopping centres: Brent Cross and Wood Green. To do so, the project draws on a combination of qualitative and quantitative methods including survey, focus group and ethnographic research. Previous studies have tended to be either highly abstract and theoretical (e.g. Fiske 1989; Shields 1992a) or thoroughly applied and technical (e.g. Retail Planning Associates 1978; McGoldrick and Thompson 1992). Apart from narrowly behavioural studies (e.g. Philips and Bradshaw 1993), little research has been carried out on people's everyday views on shopping, relating their changing attitudes and identities to the actual use they make of particular consumption spaces and places. Unlike the many studies of spectacular 'mega malls' and sites of 'heroic consumption', our approach aims to engage directly with the views of 'ordinary people' in everyday places.

This chapter discusses the variety of research methods that we used and the lessons we have learnt. The order of the chapter departs slightly from the sequence of methods that we used in the field, starting with the most extensive method of survey research, going on to discuss the focus groups and ending with the ethnography. In practice, we divided the focus groups into two phases, one of which preceded (and helped shape) the survey; the other of which came after the survey (which in turn helped inform the location and content of the ethnography). Our approach aims to combine the strengths of each method while avoiding its pitfalls. Issues that emerged from one form of enquiry were followed through into the next phase of the research. We started with a preliminary round of focus groups in Wood Green because of the more local catchment area of Wood Green Shopping City. Next, the survey allowed us to identify the wider area from which Brent Cross shoppers are drawn which in turn helped us to recruit focus groups from these areas. The survey also allowed us to identify the areas of overlap where residents were potentially able to visit either centre. The ethnography was then located in one of these areas. The combination of methods also provided an opportunity to 'triangulate' our results (Miles and Huberman 1984), increasing the validity and reliability of our results and reducing the risk of serious bias (Sykes 1991; Denzin 1993).

The questionnaire survey

The survey was a necessary foundation for the research, providing up-to-date comparative data from the two shopping centres. It also provided criteria for the selection of the second round of focus groups and helped to designate a suitable area for the ethnography. The questionnaire was designed to provide an accurate profile of people that shopped at Brent Cross and in Wood Green Shopping City and some preliminary attitudinal data that could be followed up in the focus groups. The shopping centres had undertaken their own surveys but this had been two years before the study in the case of Wood Green and five years before for Brent Cross. This previous research had been highly commercial in its orientation. Although we needed to collect a certain amount of commercially relevant information, we also wished to focus on aspects of shoppers' identity and experience that had not been covered in previous surveys. Questions therefore included the age, occupation, religion, ethnic origin and educational qualifications of respondents, the other people who lived in their household, their personal and household income, and the newspapers that they read. As we wished to compare our results with other sources, we could not make any radical departures from previous indicators of class, ethnicity and household characteristics as provided by the Office of Population, Censuses and Surveys. But we were able to add our own more interpretive questions that probed aspects of the relationship between consumption and identity that was our central focus.

For example, the notion of shopping as a social practice was examined in detail. We were interested in knowing whether shopping was a social activity carried out with family or friends, or whether it was something that people preferred to do on their own. This issue arose from discussion in the first phase of focus groups where we found that many of the mothers we talked to preferred to shop on their own or with friends rather than with their partners and children. This provided a contrast with Miller's previous experience of research in Trinidad, where the social nature of shopping in malls emerged much more clearly (Miller 1997: 293–300). Information of a more qualitative kind was collected through the use of open-ended questions investigating people's opinions of the shopping centres and on shopping in general. Participants were also asked to complete checklists of the words they would choose to describe the shopping centres and of the possible activities that they could have done whilst at the shopping centres. Finally, the questionnaire was used to identify a neighbourhood for the ethnography. Shoppers were asked to provide us with their full postcodes and this was used to construct a simple

Geographic Information System depicting the residential address of Brent Cross and Wood Green shoppers. The points of overlap where people went to either centre were places where the ethnography might take place including the 'Jay Road' neighbourhood that was eventually chosen (see below).

The questionnaire was piloted at both shopping centres and, after some initial changes, further tested at one of the centres (Wood Green). Two work-placement students from Huddersfield University (Zoe Phillipedes and Geoff Southall) were trained to help complete the survey with us for three months as part of their undergraduate training. The data were input via Quattro Pro and Excel spreadsheets and analysed using SPSS. The survey was conducted mainly on weekdays during June and July 1994, at lunchtimes and late after-noons, with some interviewing at each centre on Saturdays and weekday mornings. We maintained comparability by ensuring that the times of inter-views were the same at both centres.

The survey was completed over a 12-week period as shoppers left the centres, concentrating on the main exits but also including some interviews at less busy exits. We did not attempt to secure a quota or random sample but asked every person who passed by, and who did not obviously look in the other direction or change their path, to complete a questionnaire. Response rates are there-fore hard to assess and varied on a daily basis. On some days it would only take about two hours to complete ten questionnaires; on other days it could take four hours to get ten or fewer completed. Similar methods have been employed successfully elsewhere (Bloch et al. 1991; Feinberg et al. 1989) and our methods are broadly comparable with previous surveys of Brent Cross and Wood Green.

In order to confirm the representativeness of our survey, various checks were employed. Survey results were compared with 1991 Census data for the areas immediately surrounding the two centres. They were also compared with a 1988 survey of Brent Cross Shopping Centre by its management company Donaldsons and with a market research survey of Wood Green Shopping City undertaken in 1992 by David Peake Associates. We also sent copies of the preliminary results to the shopping centre managers and discussed the results with them to see if our results corresponded with their personal knowledge of the shoppers that used their centres.

Although questionnaires allow for more systematic comparison of data, they are, like all other methodologies, subject to bias. In our survey we found that the method favoured people with time to spare. Those who shopped at Brent Cross by car may be under-represented as two of the most-used exits to the car parks are through the two anchor stores and we were not permitted to carry out the survey there. Pedestrian counts and other forms of observational data were also collected at the two centres in order to corroborate the survey results and to indicate any weaknesses in the data. (For example, our obser-vational data clearly bear out the popularity of shopping individually or with friends rather than with other family members.)

The following discussion presents a brief summary of the profile of shoppers at the two centres, based on our survey and observational data. (For a more complete discussion of this material, see Holbrook and Jackson 1996a.) A total of 601 questionnaires were completed at the two centres, 315 (52 per cent) at Brent Cross and 286 (48 per cent) at Wood Green. The majority of respondents were shopping on their own (85 per cent at Brent Cross and 84 per cent at Wood Green) with a smaller percentage shopping in groups (15 per cent at Brent Cross, 16 per cent at Wood Green). This was partly a result of differential response rates, with some groups (e.g. mothers with young children) more reluctant than others to complete the questionnaire (which took ten minutes or more).

Public transport was the dominant mode of travel to both centres, with slightly more people (21 per cent) using private transport to reach Brent Cross than Wood Green (18.5 per cent) (see Table 3.1). The mean amount of money spent at Brent Cross was almost twice as high as at Wood Green (£31.42 compared to £16.00), with much greater variation of expenditure at Brent Cross than at Wood Green. As might be expected from its out-of-town location and 'regional' clientele, people spent longer at Brent Cross, averaging almost one and a half hours, compared to just over one hour at Wood Green (see Table 3.2). The survey results suggest that the two centres appeal to rather different age groups, Brent Cross attracting an older clientele than Wood Green.

Table 3.1 Travel to Brent Cross and Wood Green shopping centres

Percentage travelling by:	Brent Cross	Wood Green
Bus	72.7	52.1
Car	21.0	18.5
Underground	3.5	12.9
Foot	1.9	13.6
Other	0.6	2.8
No response	0.3	0.0
Size of sample	315	286

Source: Survey results

Table 3.2 Duration of visit to Brent Cross and Wood Green shopping centres

	Brent Cross	Wood Green
Mean (in minutes)	81.1	62.2
Percentage of respondents spending:		
< 30 minutes	21.8	41.5
30–59 minutes	29.4	27.3
1–2 hours	35.9	23.8
> 2 hours	12.8	7.5

The survey revealed that Brent Cross has a surprisingly 'local' character in terms of the frequency with which shoppers visit the centre. In fact, a higher proportion of our respondents shop in Brent Cross on a daily basis (9 per cent) than in Wood Green (7 per cent). The age and gender breakdown for these frequent shoppers is also significant, with more men (especially older men) shopping on a daily basis than women (see Table 3.3). These figures are supported by people's perceptions of the two centres with almost exactly the same proportion of people describing the two centres as 'local' (43.5 per cent at Brent Cross, 43.7 per cent at Wood Green). These findings raise interesting questions about the 'maturing' of shopping centres and suggest that further research is needed on the 'life-history' of Brent Cross and other well-established centres.

The centres were also chosen because of our expectations concerning the contrasting ethnic and religious profiles of their customer bases. From our prior knowledge of the centres, we anticipated that Brent Cross shoppers would be less ethnically diverse than Wood Green shoppers but with a higher proportion of Jewish shoppers. In fact, the findings were more complex. As anticipated, there was a higher proportion of Jewish shoppers at Brent Cross and a higher proportion of Pakistani Muslims at Wood Green. In general, too, Wood Green attracted a more diverse customer base with a similar proportion of Greek Cypriots at each centre but more Turkish people among the Wood Green sample. There was also a higher proportion of Indians at Brent Cross than at Wood Green (see Table 3.4). The survey data also suggest a declining proportion of White shoppers at Wood Green Shopping City since the 1992 survey and an increasing percentage of Black Africans.

Clear contrasts were expected between the two centres in terms of social class, both on their basis of their (inner London and suburban) locations and because of the range of shops in the two centres. In fact, the social differences were less marked than we anticipated. The nature of these differences was investigated by means of a variety of indicators concerning income, occupation, housing tenure, education and newspaper readership. This approach draws on conventional measures of social class (as reviewed in Crompton 1993) and on more subtle

Table 3.3 Frequency of visits to Brent Cross and Wood Green shopping centres by gender

	Brent Cross		Wood Green	
	Men	Women	Men	Women
Percentage visiting:				
Daily	11.9	8.3	9.9	6.0
> Once a week	14.3	22.9	16.8	21.9
Weekly	14.3	20.6	20.8	30.1
> Once a month	27.4	15.1	22.8	20.2
< Once a month	14.3	12.8	13.9	11.5
Other/no response	17.9	20.2	13.9	10.4

Table 3.4 Ethnicity and religion of respondents

	Brent Cross	Wood Green
Ethnicity (%):		
White	78.4	66.8
Black African	4.4	8.7
Black Caribbean	3.2	5.6
Indian	6.0	4.9
Pakistani	0.3	1.7
Bangladeshi	0.6	0.7
Greek (Cypriot)	1.6	2.1
Turkish	1.3	3.1
Other/no response	4.2	6.4
Religion (%):		
Anglican/Church of England	25.4	25.5
Catholic/Roman Catholic	11.9	12.8
Hindu	4.1	3.5
Muslim/Islam	3.2	6.9
Jewish	14.0	3.1
No religion	32.7	38.8
Other/no response	8.7	9.4

measures of 'cultural capital' (as suggested in Bourdieu 1984 and, in a British context, Savage et al. 1992) where differences in family background, socialisation, education and 'taste' are identified as markers of important social distinctions.

Contrary to our expectations, shoppers at Brent Cross and Wood Green did not differ dramatically in terms of social class as defined by occupation. While there were twice as many employers and managers at Brent Cross as in Wood Green, the proportion of professional workers was very similar (see Table 3.5). Both centres attracted shoppers from across the range of social classes and there is little evidence that either Brent Cross or Wood Green are guilty of the kinds of 'social exclusion' (of older and poorer consumers, for example, or those without access to a car) that critics have leveled at the newer generation of regional shopping centres (see, for example, Bromley and Thomas 1993b; Raven et al. 1995; Williams 1995).

Other measures of social class suggest more pronounced differences. Perhaps the most striking difference between the two centres is the number of owner-occupiers with only 40 per cent of shoppers at Wood Green owning their own homes compared to 67 per cent at Brent Cross. In terms of household income, Brent Cross shoppers also outstripped those in Wood Green (with 12 per cent reporting household incomes in excess of £25,000 per annum compared to less than 6 per cent in Wood Green). Figures for personal income showed similar disparities (see Table 3.6). The proportion unemployed is much higher in Wood Green (5.6 per cent) than in Brent Cross (1.6 per cent), corresponding to a higher proportion unemployed in Haringey than in Barnet.

Table 3.5 Occupation of respondents

Percentage:	Brent Cross	Wood Green
Employers and managers	5.3	2.5
Professional, self-employed	0.6	1.4
Professional, employee	5.1	3.8
Intermediate, non-manual	7.3	5.9
Junior, non-manual	14.0	5.6
Foremen and supervisors	0.0	0.7
Skilled manual	1.9	6.3
Semi-skilled manual	0.6	2.0
Personal service workers	8.3	10.5
Part-time	0.6	1.7
Own account professionals	3.2	3.1
Housewife	7.3	7.0
Student	21.3	22.7
Retired	16.8	14.0
Unemployed	1.6	5.6
School pupil	4.4	5.6
No response	1.6	1.4

In terms of educational background, the two centres are broadly comparable with a slightly higher proportion with no educational qualifications at Wood Green than at Brent Cross, but also a slightly higher proportion with higher degrees (which may be a consequence of gentrification in the Wood Green area). Newspaper readership also show some marked contrasts, with a higher proportion of *Sun* and *Daily Mail* readers at Wood Green than at Brent Cross and a higher proportion of *Telegraph* and *Times* readers at Brent Cross (see Table 3.7).

While the data on occupational class differences among consumers at the two centres are relatively muted, responses to more qualitative and open-ended

Table 3.6 Household and personal income of respondents

Percentage:	Brent Cross	Wood Green
Household income:		
No response/don't know	51.6	51.2
No income	15.9	17.8
< £10,000	5.4	12.2
£10,000–£25,000	14.9	13.2
> £25,000	12.2	5.6
Personal income:		
No response/don't know	13.3	9.2
No income	19.7	18.9
< £10,000	35.6	48.6
£10,000–£25,000	25.4	21.6
> £25,000	6.0	1.6

Table 3.7 Newspaper readership

Percentage reading:	Brent Cross	Wood Green
Sun	17.8	22.4
Daily Mirror	13.3	17.1
Guardian	12.4	11.2
Times	12.1	5.2
Daily Mail	10.2	13.3
Daily Express	7.0	5.9
Daily Telegraph	6.3	3.5
Independent	6.0	5.6
Today	1.6	2.4
No daily paper	8.6	8.0
Other/no response	4.7	5.4

questions suggest that there are more marked differences in terms of the 'image' of the centres and in terms of more subtle measure of social differentiation. Respondents drew attention to perceived differences in ethos and clientele between the centres in terms of class, wealth and affluence which may or may not have been borne out by more 'objective' measures (see Table 3.8). Of those who stated a difference between the centres, the most common response was from Wood Green shoppers who felt that people who went to Brent Cross were more wealthy, rich or affluent (20.2 per cent). The most common response from Brent Cross shoppers was also about their relative affluence or wealth compared to those who shopped at Wood Green, although in this case the percentage was much lower (14.4 per cent). The second most frequently stated difference between the centres was in terms of social class, with 9.9 per cent of Wood Green shoppers saying that Brent Cross was a more 'classy', more 'upper-class' or more 'middle-class' place. Apart from these differences, however,

Table 3.8 Words used to describe the shopping centres

Percentage describing centres as:	Brent Cross	Wood Green
Convenient	69.5	66.8
Modern	44.1	45.1
Local	43.5	43.7
Attractive	42.9	32.5
Safe	41.6	24.8
Familiar	38.4	30.8
Enjoyable	38.1	32.2
Expensive	21.0	8.8
Artificial	9.5	11.9
Soulless	5.1	7.7
Alienating	2.2	2.4
Intimidating	1.6	2.8
Unsafe	0.0	3.8

there were many similarities in terms of the words most frequently used to describe the two centres. Neither was thought to be 'soulless', 'alienating', 'intimidating' or 'unsafe' (mentioned by fewer than 10 per cent of respondents in each case). But over 20 per cent of Brent Cross shoppers thought the centre was 'expensive' as compared to only 9 per cent at Wood Green. The most common words used to describe the centres were 'convenient' (almost 70 per cent in each case), 'modern' (around 45 per cent), 'attractive' (33 per cent in Wood Green, 43 per cent in Brent Cross) and 'enjoyable' (32 per cent in Wood Green, 38 per cent in Brent Cross). While respondents at both centres tended to use the same words, in roughly the same frequency, to describe the centres, the main differences were that Brent Cross was thought to be safer but more artificial than Wood Green. 'Artificiality' may indeed be the downside to 'convenient' (easy parking, climate-controlled environment, everything under one roof). This emerged as a major theme in the focus group discussions where participants regularly complained about the lack of fresh air and natural light, a general feeling of claustrophobia and the sense of being cut off from nature (a point developed in chapter 6).

Focus groups

As mentioned above, our research actually began with a preliminary round of focus groups, prior to undertaking the questionnaire in order to get an idea of the kinds of issues that would be of interest to our respondents. We also felt that to go straight into quantitative research at the shopping centres without talking to shoppers first might have meant that we missed out on important aspects of the subject. It was important to get the survey research as focused as possible as this kind of research is expensive and time-consuming and it would not be possible, given the available resources, to return to the shopping centres to do further survey research. Starting with focus groups in the neighbourhoods around Wood Green rather than with survey work at the two centres also coincided with our theoretical intention of focusing on the local, domestic context of consumption rather than solely with the point of sale. For, as we have argued elsewhere, consumption is a social process that involves much more than an isolated act of purchase, reaching out into cycles of use and re-use in contexts well beyond the point of sale (cf. Miller 1987; Jackson 1993).

Departing from the market research tradition, we designed the focus groups to try to model the kind of everyday conversations that people might have about shopping without being prompted by us. Rather than asking specific questions, therefore, we raised a series of themes for discussion and then tried to limit our contributions to occasional prompting, steering the conversation back on course if it became seriously side-tracked. The majority of the groups were run by Beverley Holbrook, on two occasions assisted by Peter Jackson (for a full discussion of the method, see Holbrook and Jackson 1996b). Topics were as open as possible, in order not to prejudge people's responses and so that everyone's views were regarded as equally valid. We began with a list of six such themes:

- the kind of shopping that people particularly liked or disliked;
- the differences between shopping at Brent Cross and Wood Green;
- the differences between high streets, markets and shopping centres;
- whether participants felt that men and women shopped differently;
- shopping as work or leisure; and
- special types of shopping such as at Christmas or in the January sales.

As we hoped, the focus groups provided a number of unexpected insights, including the way in which individual attitudes are shaped by a group situation and by the wider social context (a theme to which we return throughout the study).

After a series of 'false starts', with advertisements in local newspapers and flyers distributed at the centres failing to produce much response, we recruited the groups from local youth and community centres. While we do not claim that the groups are representative in any statistical sense, we aimed to reflect the social diversity of the Wood Green area in terms of age, gender, class and ethnicity. As we have discussed elsewhere (Holbrook and Jackson 1996b), we feel that there are certain advantages to using such 'natural' groups where participants already know each other well and are happy to talk in each others' company. For one-off groups such as ours, we feel that these advantages far outweigh the possibility that pre-formed groups will prevent people from voicing 'unpopular' opinions (cf. Goldman et al. 1987; Stewart and Shamdasani 1992). We acknowledge, however, that our method is very different from the market research tradition (where participants are complete strangers, selected because they meet certain predetermined characteristics) and from the kind of in-depth discussion groups which meet on several occasions and where the development of a group dynamic is a central aspect of the research (cf. Burgess et al. 1988a, 1988b). Following an abortive effort to pay our focus group participants a cash sum to cover their travel expenses, it was found much more acceptable to offer them a gift voucher as a token of our gratitude. This changed the nature of the transaction from a strictly pecuniary one to one that more closely reflected the kind of social relationship we had tried to establish in the groups. We also ended each group by distributing a short questionnaire which helped us to identify individual contributions to the discussion which we had recorded on tape and later transcribed in full.

Our first round of focus groups in Wood Green was therefore based on the recruitment of people from the kinds of places where they met routinely to talk and socialise. Mother and toddlers' centres, old people's homes, job clubs and play centres were identified as places where people might have the time to join a focus group and where they would already meet socially. These places also offered facilities suitable for focus group discussions. After our earlier problems, we were surprised by the ease with which focus groups were organised using this method. The manager or convenor of the centres often helped set up the groups for us and we visited others beforehand to meet people who might be interested. Once people had met us and knew the objectives of the research there was very little difficulty in recruiting further participants. In this first phase of research, we recruited six groups including two groups of pensioners (one of which was an all-women group), two groups of teenagers (one of Greek Cypriot descent and one of mixed descent), a mother and toddlers' group (largely of Irish origin) and an English as a Second Language (ESL) group of unemployed people from South Asian and Greek Cypriot origin:

- *Devon Close Mother and Toddler Club* (working-class mothers)
- *Woodside Senior Citizens' Centre* (working-class pensioners)
- *Woodside Luncheon Club* (women pensioners)
- *Wood Green Area Youth Project* (teenagers, mixed ethnicity and gender)
- *Greek Cypriot Youth Centre* (Greek Cypriot teenagers)
- *English as a Second Language Job Club* (unemployed, mixed ethnicity)

The ESL job club was the only group whose members did not all know one another beforehand and the discussion was the least relaxed, further confirming our preference for 'natural' groups.

The reasons for conducting focus groups and the aims of our research were explained to participants. We also tried to ensure that everybody could have their say by asking people at the beginning not to speak all at once and not to interrupt other people. Beyond that, we endeavoured to maintain the spontaneity of discussion within the groups and only intervened when necessary, to end unproductive silences or to avoid continuous repetition of the same issue. The names of the participants have been changed to protect their anonymity. Extracts are given verbatim, except where a few words have been omitted (. . .) or added [. . .] as indicated.

Focus groups have been criticised because they do not produce generalisable results and because of the inadequacy of the 'sample' (cf. Stewart and Shamdasani 1992). We argue that this misunderstands the nature of 'case study' research where no claims to statistical representativity are made but where the argument depends upon the logicality rather than the typicality of the reasoning (Mitchell 1983). Focus group methods are particularly useful for generating theoretical insights which can be further explored through other methods, such as survey research. They are also useful methods for probing people's perceptions, as the method encourages the discussion of issues in a more reflexive and spontaneous manner than other methods. It is a more creative method than one-to-one interviews or survey research as the discussion evolves through a dialogic process where participants generate and exchange *common meanings* and *shared knowledge*. Common knowledge relies on the presence of shared concepts and themes that may be replicated across groups within a locality ・ such as Wood Green. It includes knowledge that has been acquired separately from the group in other social contexts but which has come to be accepted as 'local knowledge'. (Our analysis of the racialised nature of neighbourhood change in chapter 5 is a good example of this kind of process.)

We acknowledge that such methods may not be appropriate for all situations. Research with groups of comparative strangers may be better suited to situations in which one wishes to compare the views of people from different backgrounds. They may also be more appropriate for discussions of sensitive or deeply held views where group members may feel more comfortable with people they are unlikely to meet again socially (cf. Kitzinger 1994). It may also be that more articulate, well-educated and middle-class people are more likely to join such groups than relatively inarticulate, poorly educated people on low incomes. The second phase of focus groups consisted of more middle-class groups in and around Brent Cross and generated some interesting contrasts with the less articulate members of some of the Wood Green groups (especially the ESL group as noted above):

- *Brent Cross Residents' Association* (middle class, mixed gender)
- *Unitarian Church Coffee Club* (pensioners, middle class, mixed gender)
- *National Women's Register* (middle class, all women)
- *Canada Villas Youth Club* (working-class White teenagers, mixed gender)
- *St Andrew's Prampushers* (working-class mothers)
- *Jewish Women's Network* (middle-class Jewish women)

The second phase of focus groups was more geographically dispersed than the first because there is a much wider catchment area for shopping at Brent Cross than Wood Green. Therefore, focus groups were held in the immediate locality (Hendon) but also further afield within the borough of Barnet (e.g. Golders Green, Edgware, Mill Hill and Finchley). One group was even held in Middlesex (the Jewish women's group) as these women regularly shopped at Brent Cross. While we had focused on working-class and poorer people in Wood Green, we concentrated on more middle-class groups in and around Brent Cross during the second phase. It was important to include a Jewish group in the second phase because there was such a high proportion of Jewish shoppers in the survey. A degree of comparability was kept with the first phase by ensuring that a group of working-class youths, a mothers and toddlers' group and an old people's group were each included in the second phase. Most of the themes from the first phase of the research were retained but Christmas shopping was dropped as a theme and instead people were asked about shopping as a family or social activity in order to follow up on the survey results.

The nature and structure of the group discussions changed quite dramatically during the second phase of the focus groups. The experience of recruiting Brent Cross shoppers indicated that middle-class groups are often organised from their own homes (e.g. residents' associations and neighbourhood watch groups) or based at sports clubs and churches, whereas poorer people tend to go to community centres and drop-in centres. As a result, the discussion was often more formal and the degree of intimacy much less than that experienced at the other groups. The two working class groups were more comparable with the earlier groups in Wood Green.

The transcripts were analysed manually rather than using computer packages such as NUDIST or ETHNOGRAPH, listening to the tapes repeatedly and coding the data according to themes that had arisen in the survey phase of the research and introducing new themes that emerged during the analysis itself. This allowed new connections to be made, such as the link between discourses of cleanliness, dirt and pollution and discourses of 'race' and racism (discussed in chapter 5). Some sections of the transcripts are reproduced at length in the analysis that follows, attributing quotations to the individuals who contributed specific views; shorter quotations are not always attributed to named individuals but are taken as representative of the views of the group in which they were expressed (and are therefore attriibuted to the group). The analysis of the transcripts and identification of themes involved several members of the research team and was the subject of discussion at our regular group meetings.

The ethnography

Unlike focus groups and questionnaires, the concept of ethnography derives from a tradition in anthropology that has rarely been explicit or abstract in its formulation as methodology. Until recently (and the change has come mainly under pressure from government grant-giving authorities), students in anthropology were subject to little – if any – training in methodology, while books on ethnography as methodology tended to be written by practitioners of other disciplines such as sociology.

The intention in this project was to retain the rather loose sense of ethnography that is employed in anthropology. In this sense, the term designates no specific requirements but tends to imply an enquiry which involves: (1) some degree of participant observation, in this case both going shopping with people and spending time with them in their homes over the course of a year; (2) having some sense of the wider social context of the individuals concerned, in this case by using the household as immediate context and working mainly within one street as a potential community; and (3) analysing these observations within a comparative perspective, in this case with particular respect to recent fieldwork on the topic of shopping and shopping centres in Trinidad (see Miller 1997: ch. 7).

The selection of the precise area was based on the shopping centre questionnaire that was carried out during the first year of the research. The survey indicated that both Brent Cross and Wood Green were dominated by relatively localised shopping and that there were a limited number of areas in which we could expect people to be shopping at both sites. These areas could be specified through an analysis of the respondents' residential addresses (using postcode data and a simple GIS). The particular site we chose proved to be in easier reach of Wood Green than Brent Cross, especially with regard to public transport. The main access to Brent Cross was by road, but this was affected by extensive road works throughout the period of fieldwork which may well have decreased the use of Brent Cross during this time.

The selection of a particular street within this area centred on the paradoxical idea of not having any particular reason to choose it. We desired neither a wealthy nor a particularly impoverished street, nor one that was unusual in any other way. The intention was to avoid clearly marked social parameters that would define the area, in order to allow those social classifications that were most relevant to the topic of study to emerge during the course of the fieldwork. In the end, a street was selected largely because of a previously known individual living there who offered to introduce us to some other people in the area. The street (which we shall call 'Jay Road') mainly comprised two kinds of accommodation. One side of the street is dominated by two council estates, called here the 'Lark Estate' with 62 flats and 'Sparrow Court' with 32 flats. The Lark Estate, though more recently built, appears as the more run down and soulless of the two, being based on a high-rise tradition of long concrete corridors off which come nondescript flats. Sparrow Court, a low-rise estate, is less constricted and is less affected by vandalism. Neither estate is particularly run down and, although they are below national average in terms of property values, they are marketed as more desirable than many other such estates in North London.

The other side of the road consists of two main kinds of housing. There are some 40 purpose-built maisonettes. These are small two- and three-bedroom flats occupying alternative ground and first floors of a terraced complex and generally seen as transitional accommodation for families who would like to own a more complete house. They may be valued at above national house prices but only because of the location of the area in juxtaposition to more affluent parts of North London. Finally, there are other terraced (and a few semi-detached) houses which are of higher value but still not particularly large or imposing compared with nearby streets. At the end of the road is some further housing associated with a parade of shops.

The situation of the street is transitional between what would be seen as unequivocally a working-class district where house prices would be lower and a middle-class district leading onto a fashionable shopping centre, called here 'Ibis Pond', with much higher house prices. It was decided to expand the study to some of the more affluent middle-class streets which border onto Jay Road and, in particular, to include a small network of houses in what will be termed 'Owl Crescent'. The study, which is still continuing, has been carried out by two people. The principal investigator of those elements being used in this book was Daniel Miller, but he was accompanied when first meeting many of the informants by Alison Clarke who will be continuing this ethnography for a subsequent year with an emphasis on non-shopping forms of provisioning (cf. Clarke 1997).

The main technique used was to leaflet sections of the street in turn, detailing what the project was about, guaranteeing anonymity and arranging a time to call. The major limitation on this was the ethnographer's own domestic commitments which meant that he was mainly only able to work during weekday office hours. This produced a bias towards those, such as pensioners, who were not working or those, such as housewives, who tended to be in part-time employment. The middle-class households, by contrast, tended to be contacted through networks of friends and neighbours. Use was made of the groups such as the National Childbirth Trust and also a network that arose from a Tupperware Party attended by Miller.

To date, the study is based on 76 households of which 52 are in Jay Road itself. The main limit in including households were the absence of many householders during the day. Following the leafleting and approaching households, only 14 completely refused to take part. This means that although the study represents some 30 per cent of households in the street, it includes some 80 per cent of households where we located people present during the weekday who were able to take part in terms of their availability. Out of these 76 households, the involvement in 55 cases has been with women only; in 14 cases with men only and in 7 cases with both men and women. The family composition of those households where this information was entirely clear is given in Table 3.9.

It should be noted that the category 'single' includes households shared by several unrelated individuals. We have not differentiated on the basis of flat ownership. Several of those living on the council estate own their own properties but we felt the more marked differentiation remains whether one lives on a council estate or not. It will be evident that the council estates are dominated by single adults with (mainly young) children or without children (mainly pensioners), while the houses and maisonettes are dominated by nuclear families, although this does not necessarily imply that the adults are legally married.

Place of origin was not always clear from our informal conversations, where this was known, it included 9 households with persons of West Indian origin, 4 of Cypriot origin and about 14 others include persons brought up not in the UK but in countries in regions varying from South and East Asia, West

Table 3.9 Tenure and composition of households in the ethnography

	Nuclear	Parent–child	Two adult	Single
Council	7	10	3	15
Other	19	3	1	9

Africa to mainland Europe and South America. Of the remaining 49 where this could be ascertained about 8 have a Jewish member although in only one of these is both partners Jewish. This indicated a higher proportion of British-born households as compared to figures taken from the local primary school which suggest a higher proportion of immigrants. The discrepancy is partly accounted for by about 8 pensioners on the Lark Estate who were almost all English and indeed in most cases of local origin.

To state that a household has been included within the study is to gloss over a wide diversity in terms of degree of involvement. The minimum requirement is simply that a householder agreed to be interviewed about their shopping which would include the local shopping parade, shopping centres and super-markets. At the other extreme are families that we have come to know well during the course of the ethnography. Interaction included formal interviews, a less formal presence within their homes, usually with a cup of tea, and accompanying them on one or more 'events' which might comprise shopping trips or activities associated with the area of Clarke's study such as the meeting of a group supplying products for the home. Direct participant observation of shopping included sixty accompanied events ranging from around fifty minutes to six hours in duration. It was found that compared to other regions (such as Trinidad), residents were more protective of their privacy and disliked researchers 'dropping in' without warning. The main technique was therefore to use a mobile phone from a car parked in the street and, once it was agreed that a visit was convenient, the ethnographer could arrive within a minute or two. The form of interaction during shopping was based on a sense of the preferences of the individual concerned. A few shoppers preferred to continue as though on their own, being silently observed. Much more common was to treat the occasion as though shopping with a friend and engage in considerable conversation. But in all cases the style of interaction attempted to follow the preference of the particular shopper.

It should be noted that all informants were told their anonymity would be protected. In order to achieve this, some facts about the street, the area and the individuals have been altered. Although we are pretty sure informants would recognise themselves we hope they would not be recognised by others. This means there is a small deliberately fictive element to the ethnographic sections of the book. None of these are however pertinent to any of the major concerns of the project or the conclusions reached.

Conclusion

This chapter has introduced the range of methods that were employed in the study, encompassing survey, focus group and ethnographic forms of research. Combining methods allows us to draw on their comparative strengths and reduces their individual weaknesses. We have also provided an overview of the survey results which allow the more qualitative material on which the rest of the analysis mainly draws to be placed in perspective. Other methods were also used, including archival research, a review of the local press and interviews with various key informants (the shopping centre and town centre managers, representatives of the local police, etc.), but are not discussed in any further detail here. The next chapter is the first substantive one, demonstrating how we have drawn on a combination of methods to 'ground' the study of consumption and to draw out our analysis of the relationship between shopping, place and identity.

4

SHOPPING POLICY AND
SHOPPING PRACTICE

This chapter provides a bridge between the two main sections of this volume. So far we have introduced the shopping centres, and considered their political and regional context as well as summarising information about the shoppers from our surveys and focus groups. In the subsequent chapters we will examine in more detail some specific areas of articulation between the shopping sites and their consumption with an increasing emphasis upon the ethnographic material. This chapter begins in the first mode considering the macro-context of contemporary retail development and the political debates which have determined recent changes in policy. Such discussions tend to presume the context of consumption and writers on policy often project a series of desires and demands upon consumers. By introducing the perspectives of the shoppers themselves, as revealed in both focus group discussion and ethnographic observation, we will begin to challenge the presuppositions that are implicit in public discussion of shopping and reveal their distance from the perspectives that are generated through working directly with shoppers. At the same time we do not imply that this is simply a process of 'discovering' the consumer, since by using a variety of methodologies we are able to demonstrate that the very assumption that there exists a consistent unit to be termed the consumer is itself invalid. Instead we will focus upon the problems of contradiction and inconsistency that are uncovered when we consider policy form the perspective of a variety of shoppers in a variety of contexts.

The policy context

From the opening of the first American shopping mall at Country Club Plaza, Kansas City, Missouri, in 1922 to the latest 'mega mall' development, the Mall of America in Bloomington, Minnesota, opened in 1992, concerns have repeatedly been expressed about the effects of such large-scale, out-of-town retail outlets on existing town centres. Similar issues are currently being debated in Britain where Brent Cross was the first purpose-built, out-of-town regional shopping mall (opened in 1976). In both Britain and the United States, a similar process can be traced where the growth of out-of-town shopping centres has been linked to the decline of high street shopping, the demise of town centres and a variety of other urban ills. This has led to particular pressures as the trend for out-of-town shopping has increased in the 1980s. According to another recent study, over 44,000 shops (more than 30 per cent of the total) closed between 1976 and 1987 (Raven et al. 1995: 37). The process was accompanied by increasingly intense market concentration with three major companies (Sainsbury, Tesco and the Argyll group) now accounting for over 70 per cent of out-of-town retail sales. The dualism of 'town' and 'out-of-town' is, however, a simplification. The concentration of shopping centres within town centres is a particularly British phenomenon. According to a 1993 report by the British Council of Shopping Centres more than half (59 per cent) of the existing 950 managed shopping centres in Britain were located in town centres (URBED 1994: 19). Brent Cross itself is a suburban rather than an out-of-town mall and therefore provides an important site which is not easily configured in terms of dominant discourses about high street and out-of-town.

Responding to mounting pressure over the environmental, social and economic effects of out-of-town shopping centres, a House of Commons Environment Select Committee was set up. Its report, *Shopping Centres and their Future* (1993–4), led to a series of revisions to the Department of the Environment's Planning Policy Guidance, seeking to impose more stringent conditions on the building of new out-of-town shopping centres. Regional shopping centres, the guidelines now read, would not normally be considered appropriate in areas where there is unlikely to be a significant growth in population or retail expenditure; where continued investment in nearby town centres is likely to be seriously jeopardised; where there would be a loss of Green Belt, open space, countryside or high-quality agricultural land; where public transport could not adequately serve a wide population; or where the effect on the road network and on the overall level of car travel would be unacceptable (Department of the Environment 1993). Parallel changes were introduced to the Planning Policy Guidance on transport which now include the advice that development plans should aim to reduce the need to travel, especially by car (Department of the Environment 1994).

Implementing the revised Policy Planning Guidance (PPG 6) was the subject of a special report on *Vital and Viable Town Centres* undertaken by the Urban and Economic Development Group for the Department of the Environment (URBED 1994). In the foreword to that report, the then Secretary of State John Gummer repeated his belief that improving the quality of towns meant 'encouraging new shopping development to locations where it can reinforce the town centre' while also discouraging 'development on green field sites on the edge of cities' (URBED 1994: iii). The report itself was based on an analysis of population and employment trends in 32 British towns, plus reviews of shopping, leisure and transport trends. It, too, adopted a dual approach of helping local authorities improve town centres while dealing with planning applications for out-of-town developments. The report recognises the effects of a 'consumer revolution' combined with new technology, increasing female participation in the labour market, a rise in 'leisure' shopping and growing reliance on the private car, all of which is putting pressure on the continued viability of many town centres (ibid.: 3). According to the Secretary of State, the guidelines were revised 'to support local efforts to safeguard the vitality of towns and the economic viability of their retail centres in particular' (*Financial Times* 23 February 1995; see also: 'Out of town, out of favour', *Financial Times* 30 March 1994; 'Gummer rejects shopping schemes', *Financial Times* 1 February 1994). The Secretary of State claimed that the new guidelines 'hail the revival of the British High Street and the demise of the out-of-town shopping centre' (*The Times* 8 February 1994). Mr Gummer was also fond of invoking the idea of the 24-hour city or town centre with people living over businesses, a more vibrant night life, better street lighting and safer car parks, all designed to counter fears of urban decline and bolster hopes of urban revival. As attractive as these ideas may be, there is all too little discussion about how such changes might be brought about.

Recent policy changes reflect a much longer history about the role of retailing in the health and wealth of the nation, which can be traced back to Adam Smith's remarks in *The Wealth of Nations* (1776) criticising the British as 'a nation of shopkeepers' (or, more accurately, 'a nation that is governed by shopkeepers'). As such, these discussions involve many larger ethical and ideological issues. For example, much of the critical force behind the (largely condemnatory) stance taken by a recent report published by the Institute for Public Policy Research (Raven et al. 1995) is fostered by the growth of 'green' concerns over the larger consequences of economic activity for the environment. In direct contrast Williams (1995) argues that none of the negative claims made about the development of new shopping centres has been effectively demonstrated. Other reports are more equivocal in their representation of the issues and are primarily concerned with the poverty of previous research as an adequate basis for making current policy decisions (e.g. BDP Planning and OXIRM 1992; Thomas and Bromley 1993). In order to place current concerns within a wider context, this chapter focuses on three sets of issues that are commonly expressed in relation to the growth of out-of-town shopping centres: aesthetic issues, social concerns about the privatization of public space, and economic fears concerning the impact of out-of-town shopping on town centres.

Aesthetic concerns

The first set of debates concerns aesthetic considerations about the proliferation of allegedly bland, 'placeless' urban architecture (Augé 1995; Relph 1976). In the United States, the designer Victor Gruen is often attributed with the 'malling of America', in Kowinski's (1985) memorable phrase, leading to 'a striking sameness' across the country in design terms (Lord 1985: 211).

Accusations of homogeneity have persisted even as these developments have been refurbished or replaced by postmodern designs. Commenting on developments like those at Ghirardelli Square (in San Francisco) or Covent Garden (in London), one critic has charged that 'the glazed barrel vault, the atrium, and the introduction of daylight (a reaction from the artificially lit dark-mall earlier centres), the scenic lift, the food court, and the "clip/bolt on" cosmetic finishes, are all becoming familiar' (Beddington 1991: 6). We will have more to say about the significance of 'natural light' in the recent refurbishment of Brent Cross in chapter 6.

Part of the current debate about regional shopping centres in Britain is also aesthetic with widespread objections to what Patrick Wright has called 'a collection of vast metal sheds along the bypass' (*Guardian* 22 December 1994). Wright claims that John Gummer's restrictions on further out-of-town retail developments (Department of the Environment 1994) are too little, too late. He blames 'the deregulating "enterprise zone" strategy of the eighties' for the developments that have already taken place which include those at Gateshead (the Metro Centre, opened in 1986), Dudley (Merry Hill, 1989), Sheffield (Meadowhall, 1990), and Thurrock (Lakeside, 1990). Further such developments are planned at Dartford (Blue Water Park), Leeds (White Rose Centre), Bristol (Cribbs Causeway) and Manchester (Trafford Centre) and over 300 large-scale retail developments already have planning permission (amounting to an additional 9 million square feet or 836,000 square metres). According to researchers from the Oxford Institute of Retail Management, there were 54 applications for regional shopping centres between 1982 and 1991 (Reynolds and Howard 1993), suggesting that there has been no fall in demand for such developments. Fewer applications during the 1990s are a reflection of government policy rather than market demand *per se*.

The aesthetic argument is often bolstered by an argument about choice. Thus the House of Commons Environment Select Committee report on *Shopping Centres and their Future* insisted on the need for 'access to a wide range and choice of convenient and attractive shopping facilities selling goods at competitive prices' (1994: paragraph 6). The same concerns are frequently expressed in the right-wing press. One recent example began with the assertion that 'little old welcoming towns' with 'small, nice shops' are being 'strangled by the superstores' (*Daily Telegraph* 1 January 1994). According to the *Telegraph*'s snap survey of popular opinion, readers from Hampshire to Hertfordshire, Surrey to the West Midlands, Gloucestershire to Essex were fed up with the social and environmental costs associated with 'hatchback shopping'. What most readers wanted, the newspaper claimed, was a choice of where to shop, town or country, rather than the current take-it-or-leave-it situation. The *Guardian* columnist Suzanne Moore provides a withering critique of this model of 'consumer choice': 'The feelgood factor cannot be reduced to consumer durables ... In the heady days of the eighties ... the talk was of freedom and infinite choice, grand abstract concepts. The reality was a Sock

Shop in every railway station, a Next shop on every high street and the homogenisation of retailing which made every town centre interchangeable' (9 February 1995). Writing from her home in the inner London borough of Hackney, she continues: 'Should I need to buy another TV or even just a hairdryer I guess I'd have to drive miles to some godforsaken warehouse somewhere on the North Circular.' She concludes that shopping malls are dull, lifeless places: 'Stuffed full of bored security guards, surveillance cameras, slow fast-food, these strange, weatherless, environments have become so many end zones . . . The buzz and hum of a market-place is often absent in such a sanitised space.' These concerns are taken up by a second set of commentators who blame the problem on the privatisation of formerly public space.

The privatisation of public space

The second set of criticisms focuses on the transformation of public open spaces into privatised indoor space. Whereas the traditional town centre has a variety of open spaces, including the streets themselves, subject to a diversity of social uses and under public ownership, shopping malls restrict the uses that are permitted within their semi-public spaces. Their indoor streets and food courts are all privately owned and patrolled by private security staff. Attempts to use their open spaces for political or religious purposes are severely restricted and those who infringe the rules are subject to eviction. The public also, knowingly or otherwise, surrender certain other civil liberties when they enter the privatised space of the mall. Drawing on Canadian evidence, Hopkins (1994) describes the conditions of access and laws of trespass that govern entry to these commodified spaces. Failure to leave on the request of the management is punishable by fine (C$100 in Alberta, C$1,000 in Ontario). Failure to pay the fine may result in imprisonment. Hopkins charts the activities of developers, managers and private security firms, sometimes in co-operation with municipal police departments.

Similar issues arise in British cities where the use of closed-circuit television is being extended from the privatised spaces of shopping malls to the public spaces of town centres. Even in these more public urban spaces, the control of surveillance systems is often privately contracted and subject to only the most cursory form of public accountability (for a review of recent British evidence, see Fyfe and Bannister 1995). Our evidence from Brent Cross and Wood Green suggests that the public are rarely aware of these wider issues of access and control, welcoming security measures as a means of enhancing their safety and general sense of well-being. Indeed the centre manager at Brent Cross commented to us that after the Jamie Bulger case, where a toddler was taken from a shopping centre by older children who then killed him, the pressure from the public was to replace the previously discreet surveillance systems with highly visible cameras and equipment. Fear of the streets and other open spaces is now a commonplace feature of British cities. A recent survey on 'managing the risk to safe shopping', reported in The *Guardian* (8 November 1995), suggests that 'Shoppers are deserting the high street for shopping malls because they fear crime and feel threatened by beggars, drunks and vagrants.' Basing their conclusions on a survey of 622 shoppers in six towns and cities, the authors of the report concluded that six times as much crime was suffered by shoppers in town centres as in shopping centres (Beck and Willis 1995). The report found that town centre managers were three times more likely to consider crime and nuisance an endemic problem than shopping centre managers.[1] Shopping centre managers were most concerned about 'threatening youths', an issue to which we return in chapter 5 where the racialisation of urban fear is a significant issue.

The debate about the privatisation of public space in shopping malls is part of a much wider argument that social historians such as Richard Sennett have described as *The Fall of Public Man* (1977). Sennett's concern for the design and social life of cities echoes Jane Jacobs' argument in *The Death and Life of Great American Cities* (1961), that urban culture thrives on diversity and disorder and that controlled environments like shopping malls are responsible for eroding the vitality and dynamism of the city and particularly its spontaneous street life. Out-of-town shopping malls were, of course, partly a response to urban residents' fear of city centres and their residents which stimulated a process of 'white flight' to the suburbs and beyond during the 1960s. The experience in Britain is a little different, though urban riots and civil disorders during the 1980s deepened an existing sense of anti-urbanism among many public officials and reaffirmed the popular connection between 'race' and urban incivility (Keith 1993).

Concerns over the decline of the 'public city' can lead to an exaggeration of the extent to which city streets and public spaces were ever equally open to all citizens. As Marshall Berman remarks, 'Greek agoras, Italian piazzas, Parisian boulevards ... were turbulent places, and needed large police forces on hand to keep the seething forces from exploding' (Berman 1986: 481). None the less, critics such as Berman have made an impassioned plea for the retention (and revival) of open, democratic, public space in cities, fearing their loss in the privatised spaces of shopping malls and similar developments where 'the human mix ... is overwhelmingly White, affluent, and clean-cut' (ibid.: 483). Berman yearns for places where there is 'a thickness and intensity of human feelings, a clash and interfusion of needs and desires and ideas' (ibid.: 484), recognising that such places involve risks and have economic costs as well as social benefits.

Elsewhere, Berman has traced back the demise of public space to the 'modernization' of the city in transformations such as those carried out by Baron Haussmann in Paris during the nineteenth century. The transformation of urban space continued into the twentieth century with architects such as Le Corbusier having an extremely ambiguous attitude towards the public space of city streets. More diverse forms of street life filled him with dread and led him to claim in 1929 that 'We must kill the streets' (quoted in Berman 1988: 167–8). Berman continues: 'For most of our century, urban spaces have been systematically designed and organized to ensure that collisions and confrontations will not take place here' (ibid.: 165). Yet, in so many recent accounts of that archetypal 'public man', the *flâneur*, the stroller who populated the streets of Baudelaire's Paris, only rarely is it acknowledged that such pleasures were not available to most women or to men who were not members of the bourgeoisie (Wilson 1992). As feminist historians such as Christine Stansell (1986) and Judith Walkowitz (1992) have shown, the public spaces of the nineteenth-century city were heavily policed and subject to increasing surveillance, with women and the working classes particularly subject to regulation and control. The city and its public spaces have always been unevenly available to men and women, and the 'geography of women's fear' (Valentine 1989) is a characteristic of most urban environments.

Recent accounts of the privatisation of public space have sometimes been guilty of romanticising the civility of past cities. Geoff Mulgan and Ken Walpole exemplify this kind of thinking, writing about the 'profound change' that has overcome public space: 'Only half the population now dares go out after dark, fewer than a third of children are allowed to walk to school, and public fear of strangers regularly erupts after such public murders as Jamie Bulger and Rachel Nickell' (*Guardian* 17 November 1994). Whether or not such fears are justified by current levels of victimisation, not all the blame can be laid at the door of 'the vast shopping malls of the eighties' that 'have failed as communal spaces'. Mulgan and Walpole argue that the commercial spaces of shopping malls have failed to become a natural part of everyday life, describing them as an anachronism, like the previous generation's tower blocks. The reason for this perceived failure is that they are 'lifeless, homogeneous; too controlled to mesh in with the memories, the dreams and ghosts which fill real public spaces'.

The debate is a familiar one, especially in North America, where the contrast between private affluence and public squalor is even more stark and where Mike Davis's description of 'Fortress LA' is a chilling vision of the future of increasingly privatised urban space (Davis 1990). In cities such as Los Angeles, Marshall Berman's warning of middle-class anxieties and hysterical fears leading to the transformation of urban space into 'reservations for the rich' (Berman 1986: 484) may seem justified, while Michael Walzer worries about 'the de-urbanized wastelands we have created' (Walzer 1986: 470) as tolerance for diversity has been replaced by more sanitised and exclusionary spaces, such as the contemporary shopping mall. For Richard Sennett, urban spaces have been reduced to carefully orchestrated consumption sites which trivialise the urban experience and grossly underestimate its potential to express the complexities of how people might actually live (Sennett 1990: xi–xii).

A variant of the privatisation thesis is the argument that shopping centres are socially divisive, restricting entry to those who have access to private cars and excluding those who lack the necessary cultural or economic capital, who are marked out as 'undesirable' because of their appearance or 'threatening' behaviour, or simply because they are 'loitering' (not spending money). The evidence of social exclusion is contested, with some observers claiming that access to private transport has no significant effect on customer profiles, while others argue that the clientele of regional shopping centres such as Meadowhall are unduly biased towards the affluent (Howard and Davies 1992; Rowley 1993). There are, however, clear differentials in some of the costs associated with town centre and out-of-town shopping. For example, in inner London, 60 per cent of trips to a Sainsbury superstore were made by car, compared to only 8 per cent of trips to a nearby high street store from a similar catchment area; while in Newcastle, 27 per cent of town centre shoppers travelled by car compared to almost 80 per cent of those shopping at the out-of-town Metro Centre (Raven et al. 1995: 10). What is sometimes forgotten is that one of the early arguments for the development of such centres was that they helped relieve traffic congestion in the inner cities. These arguments take us to our third theme: assessing the economic effects of out-of-town shopping centres.

Economic effects

A third set of fears centres on the economic consequences of out-of-town developments. Again, the evidence is disputed. Some have maintained that there is a direct relationship between the growth of regional shopping centres and declining employment in town centres; others suggest that regional shopping centres create jobs; others still that they are merely transferred from one place to another (Howard 1989; Whitehead 1993). Reviewing the evidence, a recent Institute for Public Policy Research report concludes that 'out-of-town retail development undermines the viability of established town-centre shopping facilities' (Raven et al. 1995: 26). The evidence of declining retail sales in many traditional high streets is clear but the causes are disputed. According to some observers, planning trends, shifting consumer tastes and the effects of economic recession may be as much or more to blame as the development of shopping malls (*Economist* 1 February 1992). Writing about the contribution of shopping centres to the decline of cities, David Blunkett (Labour MP for Sheffield–Brightside) argued that 'once the core of a city has died, as can be shown from North America, it is very difficult indeed to retain any sense of community or promote other programmes for economic or social regeneration' (letter to the *Independent* 29 February 1992). He argues that the city centre has been denied the infrastructure investment available to the Sheffield Development Corporation and to the developers of Meadowhall, while advocating the appointment of town centre managers, such as those at Wood Green and thirty other local authorities. Through its 'basic credits' approval policy, the government restricts the amount that public authorities can spend on city centres while no such restrictions are placed on private operators who can borrow and invest unlimited amounts on shopping mall developments. The Labour Party's response has been to try to link the future of shopping centres to transport policy, seeking to balance the popularity of private car travel with increased investment in public transport.

There is some support for this view from the retail industry itself. Stuart Hampson, chair of the John Lewis Partnership, warns that limiting out-of-town developments will not in itself guarantee the revitalisation of town centres. Making town centres attractive, he suggests, requires better public transport and car parking, and public investment, with 'a clear and consistent (policy) framework in relation to planning, transport and infrastructure' (*Financial Times* 23 February 1995). Other spokesmen for the British retailing industry are similarly frank. The chief executive of the Institute of Grocery Distribution warned in 1993 that 'old people and the disenchanted poor ... will have serious problems with where to buy food because of the growth of superstores and the lack of town centre stores' (quoted in Raven et al. 1995: 35).

Despite the former Conservative government's insistence on the need for an unfettered market, the regulatory environment has always exerted a vital influence on the British retailing industry. From the ending of Retail Price Maintenance in 1964, through the Fair Trading Act (1973), the Restrictive Trade Practices Act (1976), the Resale Prices Act (1976) and the Competition Act (1980) to the liberalisation of Sunday shopping in the 1990s, government has always played a crucial role in shaping the retail environment. Given the proliferation of legislative and other policy measures, it is hard to sustain the traditional image of the supermarket as an epitome of the 'free market', driving down prices and increasing consumer choice. The Institute for Public Policy Research (Raven et al. 1995) challenges the industry's traditional image of economic efficiency and competitiveness. Reviewing the evidence of increasing transport costs and other environmental impacts which are not paid for fully by the retail industry, it shows that increased choice for the most affluent has led to decreased choice and higher prices for poorer members of society including those without access to a car. Our own survey results, however, confirm that questions of convenience, safety, cleanliness and choice remain high on the list of reasons cited by consumers in choosing to shop in places such as Brent Cross and Wood Green.

To conclude, it is clear that discussions about the future of shopping are part of a much wider debate about the future of the city, the role of the state, the nature and aesthetics of public space and the regulation of the 'free market'. Since the empirical evidence is so patchy and contested, critics have suggested that the current debate about out-of-town shopping is better approached as a moral question rather than as a debate about more narrowly defined economic issues. As our own research shows (see below), apparently straightforward issues of 'convenience' or 'choice' are rarely matters of simple utility. As we argue throughout the book, shopping involves the articulation of complex social distinctions and raises significant challenges for simplistic conceptions of 'identity'. So far we have sought to sketch the contemporary policy context which shapes people's experience of Brent Cross and Wood Green shopping centres. We now turn to the perspectives of the shoppers themselves.

The shoppers' perspective

What is a shopper's perspective?

In almost all of these debates about the future of shops and shopping we find a variety of key concerns projected onto actual shoppers. In some cases there has been a process of consultation, but often and, in particular with the more academic writings, there is rather an assumption made about shoppers' concerns and tastes without much self-doubt that the opinion of the writer would be shared by the population at large. This may often be based on the author 'reading off' from the built environment to construct their interpretation. To pre-empt our conclusion, the most conspicuous problem with the public debate about the future of retailing is that it tends to consist of a middle-class projection onto the poorest and most constrained sections of the population about what would be best for that population. In general it misrepresents that population, since while it is possible to find evidence that accords with these projections, there is alternative evidence to suggest that many of those so designated have quite the opposite opinions to those assumed.

The situation is, however, rather more complex, since it will be argued that there are at least three points from which one could extract something called 'an opinion' and these are by no means consistent. These are: what people say in public, what people say in private, and what they actually do irrespective of what they say. 'Public opinion' turns out to be as much a comment upon the context of enunciation as they are on the topic they purport to be about. This may be in part because people feel the need to accord with what is thought to be the appropriate expression of value expected of them by others. This normative pressure may be just as important a factor in accounting for what they say about shopping as those of comfort and convenience.

In order to explore these issues in more depth a single case study will be presented, with the implication that other people, if considered at the same level of detail, would reveal parallel issues. The elderly and especially elderly single-person households are a rising segment of the population. Some 20 per cent of households in Britain are now occupied by persons over 65 years old. They are often those who suffer most from the practical difficulties involved in the very logistics of shopping and these issues are often raised in public discussion of the future of retail. Most of the ethnographic work on the elderly was based in lower income groups where these difficulties were particularly evident. Indeed there was often a sense within the ethnography that disability, poverty and loneliness have become closely correlated with large groups of elderly persons that have been in effect stranded in council estates with diminishing access to welfare assistance. The evidence certainly supported those in the criticism of retail development who argue that this represents a key group whose concerns have often been neglected.

Our own attempts to gauge the opinion of this group fell into two main categories: the focus groups and the ethnography. We convened two focus groups which involved elderly males and females respectively. In both cases the conversation starts with discussion about quite practical issues such as the state of the toilets. What emerges fairly early on, and remains as a constant motif throughout both focus groups, is a general critique of modern shopping forms expressed within an evocation of nostalgia which constantly asserts the superiority of shopping several decades ago.

As an example, the conversation within the women's group contains the following exchange:

> Nice quality shops used to be in Wood Green High Road.

> It's still all foreign muck, isn't it – that's the way I look at it anyway. I mean even the cameras – look at the trouble I've had with cameras. I says: 'Good God, another one with a flash', Hong Kong muck. It's true, I mean I took it back and told them, but they wouldn't give my money back, they give me another camera. But it's right, isn't it?

> Yeah, well the store, Sainsbury's – now we used to have a Sainsbury's in Wood Green years ago – three shops it used to be, in the main High Road, and I don't know about other people, but we used to deal there every week and there used to be queues, but somehow or other you. . . .

> Not like they are now.

> No, you used to get the service. I mean they used to cut you the size of cheese you wanted, you know, or the butter.

> Another thing, you go into any of the supermarkets one day to get your goods – I bet if you go the next day or perhaps the next week, you'll not find the goods that you want in that spot.

The men's discussion is not quite so dominated by this litany of complaints but it also has many instances where it moves into a comparable nostalgia:

> Shopkeeping is totally different today, it's all big stores and they represent money whereas years ago if you saw someone you had a customer and if he came in and wanted something else you'd tell him where he could go and get it.

> I think Jim's talking about the days when if you went to a grocer's shop or something and gave him a pound note he'd say 'I can't change that' – he'd most probably took about 4 or 5 shillings you know that's the sort

of thing. I mean those were the days, that can be recorded, but those were the days when you'd get all your weekend shopping and change out of 10 shillings, 50p how about that. I was telling you last week.

The good old days.

There is then plenty of material within these focus group discussions to suggest that the elderly are highly dissatisfied with contemporary shopping and find a wide range of lamentable changes from former local corner shops. If these discussions had been our sole source of information then it would be reasonable to claim that the views of pensioners were more or less consistent with the way they are represented in public debate. Even this would depend on a rather partial representation since there is alternative evidence within these same focus groups that would make this appropriation of their views problematic. For example, although the corner shop is evoked as the symbol of a time when one could measure one's purchases in a few shillings, it does not follow that this is true of the contemporary corner shop. On the contrary, it is clearly noted that these have now become too expensive for most of the elderly to be able to afford to use them.

This judgement is, of course, relative. Any historical account of the cost of living would show that the elderly are actually mistaken in their belief that most things were once cheaper, and the overall proportion of budgets being spent on food has clearly declined. The sense of the expense of the current corner shop is actually based on the comparison with the larger supermarkets which today are almost invariably cheaper for most basic commodities (Raven et al. 1995: 29; for the larger context of changes in expenditure see Benson 1994: 11–55).

Not only are these opinions relative to different shopping experiences but they are also relative to the position of this group within the life cycle. These speakers are mainly living entirely upon their pensions and are therefore objectively poorer today than at an earlier time when they were in work. This would have the effect of making goods appear more expensive or at least less affordable than they used to be. More important still as an influence over what is being said may be the context of the focus group itself. At a small scale much of what is said within the focus group appears to work as the construction of a community based around the common experience of age, where almost any reference to the past evokes supportive and empathetic responses from the others based around a collective desire to believe in what is explicitly termed 'the good old days'. One can almost sense the contradiction in the quotation given above when one of the women is about to admit to the presence of queues in the old days but checks herself with the phrase 'but somehow or other you . . .', which is immediately taken up and supported by another speaker who affirms by implication that a queue in the old days couldn't possibly be compared with a queue today. This is a not unreasonable claim given that they may now be a great deal more frail.

These conversations cannot be treated as clear expressions of opinion about the contemporary shopping experience. They are taken from a context which is as much about developing a common ideology based around age, in which almost any area of modern life is going to be subject to a rigorous critique from the perspective of a positive nostalgia. It is therefore not that surprising that when we turn to the private lives and views of the elderly as uncovered during ethnographic study the picture is extremely different.

Mr Jones will serve as an example. A widower in his eighties, he does not find it at all easy to travel to the shops, although he has been the major shopper in his household, not only for the seven years since his wife died, but even before that as she was ill for quite some time. He has two children living in Essex but they rarely visit or help with practical matters. He recognises, however, that he is much better off in this respect than his neighbour, a slightly younger woman, who is severely disabled by her reliance on frequent specialist medical care which is based around a machine in her apartment which helps with her breathing. She is completely unable to shop and Mr Jones is her main support in carrying out regular shopping on her behalf, though others on the estate also help out from time to time. On the occasion when he was accompanied she had given him a written list for two loaves of thick white bread, one pound of butter, Red Label tea bags, one box of Shredded Wheat, one tin of ham, one garlic pâté, one blackcurrant jam, one box of beef burgers and two pounds of sugar.

Mr Jones always shops in Wood Green, usually taking the bus. Given his age, he can only carry two bags of groceries, but like many of the elderly he finds he must carry two rather than one, because he cannot keep his balance when carrying a single shopping bag. It would clearly be in his interest to shop from the local shops on the street and most of the standard goods that he buys might be available there. Nevertheless he is adamant that he would not dream of using local shops, or only as an absolute last resort. This is partly based on cost, but also a preference for the range and atmosphere represented by the supermarkets. In private, Mr Jones expresses a marked preference for the larger supermarkets, such as the Sainsbury's or Kwiksave in Wood Green. But on the other hand he finds some supermarkets such as Safeway in Wood Green too large, since he feels that he might get lost there.

If we expand from this single example to that of all the elderly who took part in the ethnography then it may be said that while some of them do express in private the same kinds of fear of massive spaces and becoming lost that tend to dominate their public discussion of these issues, these are clearly outweighed by what they see as the major advantages of these larger sites. There are a wide variety of factors that result in this preference. One factor for most, though not all, of the elderly is racism, in that the positive feelings towards those who worked in the corner shop during their youth has been replaced by negative feelings towards the Asian shopkeepers who run most of the local shops. Equally important is the price differential between corner shops and supermarkets. The elderly have become amongst the most divided sections of the population in terms of income, between those with accumulated savings from middle-class occupations and those dependent upon the state pension alone and suffering from increasing cutbacks in associated benefits. The elderly included in this project were in some cases the most impoverished and destitute of all those involved within the ethnography. Clearly, saving money should be a major factor for them.

At this point we might be tempted to contrast thrift as a practical factor for the impoverished as against racism which might be termed an element of ideological construction. But even this is not straightforward. Many of the pensioners have an attitude to thrift which raises it well beyond the mere practice of a necessary skill. The discussion of price and the claims made to money saved demonstrate that thrift is part of a much wider system of valuation and valorisation related to the ideals of traditional housewifery that emerged as a practice when the same individuals were much less impoverished than they are at present. Furthermore, in many cases, the money saved is not used to further their own interests, but to buy what might be seen as extravagant presents for descendants who fail to reciprocate in terms of affection or attention returned (for a fuller discussion of the place of thrift in shopping see Miller forthcoming). The ethnography demonstrates that even in conditions of relative destitution saving money turns out to have surprisingly little to do with pragmatics and need, and is rather a core strategy in the construction and maintenance of ideals and values.

But the biggest problem in coming to understand the perspective of the elderly lies in the degree to which other people project upon them certain values and tastes which are inappropriate. In short, it is often assumed that the elderly are largely a nostalgic and sentimental community whose ties lie firmly attached to the traditions of their own childhood, and that they would therefore ideally prefer a sepia-tinted image of 'olde-worlde' shopping. The ubiquity of this assumption in discussions of retail has been noted as part of what is probably the most thorough summary of research to date: 'frequently the assumption is stated over and over again that elderly people prefer more traditional forms of shopping. Typically it is also supposed that they prefer to shop closer to home because they find travelling difficult or expensive. However, there is very little empirical evidence to back up this assumption' (BDP Planning and OXIRM 1992: 77).

One can easily see where this assumption comes from. It was confirmed by the nostalgic mode of focus group discussions. Furthermore, compared to most people, pensioners are often surrounded by objects which do indeed possess a patina of memories and incorporate a history of gifts or other social relations. When, however, we turn to opinions given in private and to their actual practice, then the elderly emerge as far more unadulterated modernisers than their descendants, with a strong preference for what they regard as the most modern and contemporary forms as against the kinds of styles they have been trying to repudiate for much of their lives. In some previous work on kitchens (Miller 1988b) one of us had already found that it was the youngest informants who preferred the 'olde-worlde' style of fake oaks, the middle generation who led the move to stripped pine which represented a kind of compromise between modernity and an evocation of tradition, while it was the elderly who most positively associated with modernist and functionalist styles and plastics, such as melamine, laminates and vinyls.

With respect to shopping, the elderly commonly showed a marked preference for the large supermarkets precisely because of their desire for bright modern looks and access to the large gamut of new produce. Even where they may be conservative in actually trying out such new items they like to be associated with them and be aware of these innovations. Their individual (as against their collective) memory of traditional forms of shopping is often related to rationing, lack of choice and in general poor quality in the sense of rotten or diseased food items. The other advantage of the larger shops is the general feeling of security and safety and ease of access compared to earlier forms of shopping facility. While some pensioners make these preferences clear in conversation with others it only emerges through observing the choices they make in practice.

While Mr Jones may feel the Safeway in Wood Green is too big, Mrs Poole uses this as what she calls a 'small shop' for forgotten items in contrast to her major weekly shop which is at an immense Tesco outside the local shopping centres. Mrs Poole knows Mr Jones since they are both part of the diminishing pool of people who originally settled on this council estate when it was first built. She is 76 but she always takes with her a friend aged 86. The two had become friends when they both lost their husbands at around the same time, and since Mrs Poole is far more alert and capable than her friend she assists her with shopping and other matters. Like Mr Jones she travels to Wood Green by bus, and reckons on going shopping for groceries at least three times a week, in part as a means to get away from the apartment and be with a companion. When asked about Tesco by the ethnographer (Daniel Miller), she replied as follows:

MRS POOLE: It's a very good shop, they're very obliging, if you have any problems you just go to the desk sort of thing and they'll sort you out and you don't have to queue for too long you know. They've got a terrific amount. No they're very friendly and I mean if you can't find anything, which very often happens you, because they will move things around you know, if you ask someone they just don't point you, they take you and say well here it is, sort of thing.

MILLER: Don't you mind the size?

MRS POOLE: Oh no as long as I can hang on to my friend because she's 86 and she wanders off you know, she'll say 'I'm going to get . . .' and I say no I'm coming with you! I've lost her before and it's taken about half an hour, or so it seemed, you know, 'have you seen a lady wandering around?' No I quite like Tesco's. And we have lunch there.

MILLER: Inside Tesco's?

MRS POOLE: Yes there's a nice little restaurant, well cafeteria you know and you can have a soup or they do three or four hot dishes or jacket potato salad, anything you know, and that's sort of a little outing you know sort of thing.

MILLER: Do you enjoy trying new things?

MRS POOLE: Oh yes, I mean yes we do and one time I would never have bought ready made, but now I'll think 'oh that looks nice'. I bought a lasagna and spaghetti bolognese. Well normally I would have made that myself but I thought 'oh that looks nice' and I've got them in the fridge at the moment, was going to have one today.

It is only when actually shopping with Mrs Poole that one appreciates the sense of enjoyment associated with the activity. She is constantly joking and giggling with her companion. As she passes the Galia melons she says 'they were so delicious, you can tell the good ones because they smell so awful', and when she has finished laughing she jokes with the man emptying sacks of potatoes into the display about how he ought to have extra long arms. There is a serious side to the skills shown in the assessment of price and the task of thrift, as well as sympathy for exploited workers in the store, but as a whole she can turn this massive store into an abundance of stimuli for joking and consideration. The fact that she has three subsidiary supermarkets she also visits on a regular basis gives her grounds for long discussions over comparative prices and adamant assertions about who provides the very best crusty white bread rolls.

Although these examples are mainly of supermarkets the same applies to the malls themselves. There are elderly persons who are intimidated especially on first acquaintance (an example is given in chapter 5) while others see shopping centres as about the only place where they feel secure within a leisure activity that costs nothing unless one actually chooses to purchase goods. The aggregate statistics on out-of-town shopping centres demonstrate that the elderly are amongst the lowest users of these facilities which are dominated by families with cars, but the evidence also suggests that this may reflect current inability to shop according to their preferences.[2] What almost all the policy-based research on shopping shows is that the elderly (as indeed the poor) are those most reliant upon that form of local shopping where prices are highest. What does not follow is that the corner shop or local high street would be preferred to the opportunity to use the mall or large retail development, even where prices were not a key issue.

So far we have discussed the issues raised by problematising our three methods for gauging the elderly's 'opinions': that is the focus group, the private opinion and the experience of shopping with them. But even if we were to remove this issue and take a step down to the apparently simpler and more pragmatic issue of convenience, the pensioners fail to conform to the position they have been credited with in the larger arguments about retail policy. One of the key factors is transport. This is relevant to the aesthetics of the landscape, the privatisation of public space and the economic context of retail development. But the implications of transport for the elderly are by no means straightforward. The question of distance is made far more complex by the fact that pensioners have passes which allow them free use of public transport. This means that as long as the shop concerned is on a bus route, it may actually be more convenient for them to sit on the bus until they reach the larger shops which can command such transport provision than to struggle to get to a shop which may be geographically closer but doesn't actually have a bus coming to its door. In general conversation about shopping as part of daily life the use of free public transport was a dominant motif. While there are many amongst the elderly who are unable to use public transport, there is a group of what might be termed the 'elderly but fit' who have colonised the new possibilities provided by free transport with alacrity and are becoming familiar with quite far-flung retail possibilities as they explore the networks of bus routes as a key leisure activity.

It is also assumed that since the elderly are one of the groups with less access to cars they have thereby been discriminated against by the development of car-based retail as against shopping in the high streets to which people can walk. What is quite forgotten within this argument is that the elderly are not just the group with less access to cars but often the group that has the most difficulty walking. For many, what is often quite a long walk to a high street followed by the need to walk back burdened with shopping is simply out of the question. In such cases the attempt to ban or restrict car access may actually hit the elderly (especially those who cannot use buses and trains) more than any other group. In practice, most of the elderly use a variety of techniques ranging from phoning for a minicab from one of the major supermarkets that provides this service to the use of local buses in order to overcome this problem. For the most disabled, the only option is to have other people shop for them which is less of an imposition when the shopper is including this task within their own car-based shopping trip. To conclude, when it comes to transport as with a number of other issues, the category 'the elderly' needs to be disaggregated into groups that may have quite diverse requirements depending upon their ability to walk, to use public transport or to drive cars (see also Westlake 1993).

The elderly have been used as a case study to contrast our results with those of most of the current policy-based discussion. Similar issues also arise with respect to other sections of the population. In general explicit opinions vary considerably, for example, between those who only shop in open high streets since they find enclosed centres claustrophobic to those who feel more at home within a covered domain. It is clearly the wealthier householders who patronise the smaller shops, but since the main array of such shops is in Ibis Pond and this is conspicuously more expensive than the shopping centres it is not possible to separate out this preference from a consideration of price. However, in as much as many of the North London shopping 'villages' are becoming dominated by high-priced retail outlets, a pattern is emerging by which the middle class becomes the only group able to afford these prices and patronise such centres, while the same group use the impoverished and the elderly as the main argument for the retention or expansion of such centres relative to larger retail developments.

For younger shoppers the single major factor in determining preference for one site rather than another is probably ease and security of parking, and there was constant discussion over this factor. Indeed during the ethnographic study of Wood Green the main finding was that hardly any two households went about the task in the same way. One shopper has a complex route through winding side-streets in order to find a parking space in a road that is difficult to access near the back of some shops. The next would swear by the practice of parking in Safeway and then spending a few pounds at that shop in order to avoid paying parking charges. Another pays a deposit at the Iceland car park while another uses the main shopping city car parks. For those who would wish to reduce reliance on the private car in general, the sheer prioritisation of the car over other factors by shoppers made for a depressing experience, but it is precisely here that the bulk of consumers are in effect in collusion with commerce over the prioritisation of private interests against the larger interests of the community or even the environment. This was consistent with the finding that green concerns, while a feature of conversation, were almost invisible as a factor in practice.

The situation with respect to children is even more complex, since the precise question of whether a covered shopping centre makes for easier shopping depends on the age of the child concerned. Even the most casual visitor will be aware that Brent Cross and baby buggies are closely associated. The advantages of not worrying about a toddler toddling into the road or into a rainstorm are clearly evident, although there are often complaints about the lack of lifts as against escalators or the lack of a crèche. In general parents are as aware as pensioners that it is only the very large-scale centres and shops that are likely to develop specialist facilities on their behalf. In the meantime it is clear from their current choice that the centres compare favourably with the small high street shops where there is a galaxy of problems in negotiating buggies, asking to use staff toilets for their children and in general dealing with the crowds.

The conclusions drawn from our ethnography may be compared with another recent study of shopping by sociologists concerned to compare the contemporary experience of living in two Northern cities – Sheffield and Manchester (Taylor et al. 1996: 141–7). The most relevant section of their report is that which compares the situation of shopping in Meadowhall, a large out-of-town mall, with the previous experience of shopping in Sheffield. In their case a similar conflict of ideology may be discerned. On the one hand there is clear evidence of enthusiasm for Meadowhall in their focus groups. Within this the three main sources of positive appraisal turn out to be a group of the disabled, a group of the elderly and a group of unemployed Pakistani housewives. Once again the main critique of these developments is a nostalgic discussion of local corner shop shopping. Despite all of this evidence, the authors hang on to the presumption that new shopping developments such as malls are good for the mobile middle classes and must be bad for peripheral or disadvantaged groups. They do this through highlighting the nostalgic discussion under a separate heading called 'The Local Shopping of the Poor' (ibid.: 148–9), while with respect to their own evidence for positive appraisal they use phrases such as 'However its appeal remains insidious' (ibid.: 144). On occasion they will admit to this contradiction, as: 'There are those, however, who expressed unreserved admiration for Meadowhall, and interestingly these often tended, in our research, to be older residents' (ibid.: 145).

The research reported by these authors managed to obtain the kind of positive appraisals within focus groups that tended in our work to come mainly from the private conversation of individuals and to be avoided in the more nostalgic context of group discussion. One would have thought therefore that a combination of the disabled, the elderly and unemployed Pakistani housewives was as clear a representation of the views of disadvantaged consumers as could be imagined. Nevertheless, so strong is the concern to bracket shopping malls within assumptions about shopping that dominate the literature, that when it comes to the table of types of shopping the authors summarise the use of the malls in the two words 'New *flâneurs*' (ibid.: 159), a term whose implications seems to flatly deny the evidence accumulated from the use of these focus groups. It is likely that the results of the study will be used merely to affirm rather than problematise what we have presented as the dominant discourse within current discussions. Their research also suggests that although the evidence we have presented is focused largely upon the impoverished elderly from a single ethnography, there are grounds for seeing our conclusions as having a much wider relevance to current debates. We have already noted in chapter 3 that, on the basis of our survey results, there is no evidence to support a thesis that shopping centres tend to create social exclusion.

This chapter has sounded a note of caution about making assumptions that we know how to assess the rights and wrongs of contemporary policy shifts. As has been shown in the first half of this chapter, a wide variety of interest groups have represented themselves as standing for public opinion. Sometimes it is evident that this is problematic. For example, Marsden and Wrigley (1996: 43) have noted the degree to which the state has allowed the major retail chains to project themselves as representing the interests of the consumer. The most emotive and divided issue of recent years with respect to retailing has been the debate about the expansion of out-of-town retail centres. Although the centres we have studied are better characterised as suburban, many of the issues raised in these debates apply to them. As has been noted such developments represent for some the triumph of commercial interest serving privileged consumers at the expense of the dispossessed. It is this stance which probably underlies the conclusions drawn against their own evidence by the authors of the study of Manchester and Sheffield to which we have just referred. In conclusion we wish therefore to make clear the narratives which seem to underlie both current debates and current research.

Conclusion

There are two rather simple, not to say simplistic, stories that may be told about recent debates over the development of large shopping centres or hypermarkets. The most simple and at first most attractive scenario is one in which government was seen as in cahoots with commerce in a policy that led to a wide range of changes that are considered to be indefensible and detrimental, most particularly the building of huge out-of-town retail outlets at the expense of the environment and welfare, and permitting the development of large supermarket chains. This has led to vociferous opposition (summarised in Raven et al. 1995). In response, the government, with an eye to public opinion and forthcoming elections, recently changed or even reversed their policy with respect to the future of shopping developments.

This story may be challenged by another simple story told in the second half of this chapter. The alternative narrative would assert that the impoverished and particularly the elderly are generally ignored in such policy making debates. Academics and other middle-class interests have by and large projected onto such people concerns that suit the current ideological premises of the middle class. As will be suggested in a subsequent chapter on nature there is a powerful force at work on behalf of a concept of authenticity which fuses notions of nature and of the past.

A typical middle-class opinion expressed during the ethnography was:

> I just think supermarkets are taking over everything and they're contributing vastly to small shops and things like fruit and veg stalls closing down and what you're missing then you end up having to drive to a shop which is not very good for the environment, not very good for anybody really, because all the local shops have closed down and particularly it's not good for people who haven't got a car, old people or whatever and also it completely impersonalises the shopping. I mean shopping is or used to be part of social life you know and that's completely disappearing.

This is spoken by a shopper who can afford the prices in Ibis Pond and will patronise such small shops, but who here defends her concerns through projecting needs onto other groups such as 'old people'.

In public, such ideological hegemony is unchallenged and the poor and elderly on the whole will reiterate what they are expected to say. Furthermore, these opinions are in accordance with a dominant discourse amongst the elderly based on a genre of nostalgia where everything was better in the past. But in private and in their actual behaviour there is a very different story which suggests that the development of large-scale shopping facilities may have particularly benefited the poor and those with various difficulties or special needs. If the changes currently being made to policy result in transport systems and policies which favour inner-city retail developments, then it is very likely that the one group who will be faced by higher prices and inner city traffic congestion will be that which can least afford to bear the brunt of these changes. Within the ethnography by far the most destitute segment of the population tended to be single elderly people living in council accommodation. This group would gain very little from the preservation of what is largely privately owned out-of-town green landscapes well away from their residential areas. This story is unlikely to be told since even those who would be the victims are too well aware of what they are supposed to say about current ideological movements respecting ecological and environmental concerns.

We have concentrated on presenting the evidence for this second story, because it seems to us the more hidden narrative which we have had to excavate through our research. But this is not to deny that both of these stories may be equally simplistic, and mask a much more complex and contradictory world. We are not predisposed to favour this second story, since it is those who show most concern with the elderly and the impoverished as well as larger green issues who tend to condemn new developments. By contrast it is often an amoral force of commercial interests that tends to favour them. Nevertheless, we cannot, as we have accused others of doing, ignore the existence of the counter-narrative we have uncovered.

When comparing these two stories there do not seem to be clear grounds for privileging either one. After all we cannot say that private opinions or practices are somehow more authentic than publicly based opinions . None of these things represents a true or objective condition to be called 'public opinion'. The only authenticity probably lies in admitting that people are inconsistent in their beliefs and that opinions are context dependent. If we have stressed the counter arguments to current discussions in this chapter it is because we feel it is essential that public opinion is recognised to contain such contradictions and ambivalence rather than being projected upon as the given interests and opinions of particular groups.

This point holds for much more than just shopping policy. The advantage of ethnographic observation is not that it reveals some hidden 'true' opinion, but that it provides evidence which can be used to challenge policy which too glibly asserts its foundations in something called public opinion. Research reveals the complex articulation between factors which help to determine benefits at particular times. The elderly vary considerably in their needs according to whether they can drive, or mount the platform of a bus. They also tend to want both the excitement of innovation and modernity and the familiarity and nostalgia of tradition – even if that tradition is invented. As we will find in the subsequent chapters the main findings of our detailed research have constantly to return to the evidence for contradiction in the relationship of place and identity.

5

FAMILY SHOPPING AND
THE FEAR OF OTHERS

In this chapter we argue that 'family shopping' is one of the key contexts in which the relationship between consumption and identity is currently being forged. By this we mean that the majority of shopping decisions are made in respect of a relatively narrow range of social relationships, predominantly within a familial/domestic context. Few of our respondents engaged in the kind of hedonistic, self-indulgent consumption practices that have been celebrated in the literature of 'lifestyle shopping' (as described by Shields 1992a and Featherstone 1990). Far removed from the world of luxury and personal pleasure, our respondents generally adopted a language of thrift and social responsibility (particularly in the case of families with small children). There is an important distinction, though, between *going shopping as a family* (which few of those we surveyed at Brent Cross and Wood Green actually did, and which those who did generally disliked) and the *familial context of shopping* which, we argue, is the dominant context of contemporary consumption as revealed by our focus group and ethnographic research (see also Miller forthcoming).

We also argue that the popularity of shopping malls and planned shopping centres such as Brent Cross and Wood Green can be attributed in large measure to their success in managing diversity, promoting the virtues of familiarity and reducing the risks of casual encounters with socially different others. Through the privatisation of public space, the perceived dangers of high street shopping have been 'domesticated' and the majority of middle-class consumers have been made to feel safer.[1] The perceived risks of shopping at the two centres are shown to be related to the degree of (un)familiarity with those who shop there, including 'foreigners' and others who are considered 'out of place'. Many of the narratives encountered in our focus groups (particularly those relating to fear of crime, safety and surveillance, and the nature of neighbourhood change) can be related to notions of familiarity and difference. 'Familiarity' is therefore constituted in opposition to various kinds of 'otherness', affirming the boundaries of inclusion and exclusion along lines of gender and generation and, particularly, according to distinctions of 'race' and class.

Finally, we note that the shopping centre managers are implicitly aware of these distinctions, using the metaphor of 'family shopping' in their marketing strategies to emphasise the centres' warmth, safety and convenience as places to shop. We begin, though, by establishing that 'family shopping' is a thoroughly gendered construction.

The gendered construction of 'family shopping'

The association between women and shopping has a long history. Judith Walkowitz (1992: 46–7) argues that shopping emerged as a newly elaborated female activity in the 1870s, allowing middle-class women to enter public space unchaperoned for the first time and giving rise to all kinds of anxieties about their changing social roles. Gillian Swanson (1995) suggests that the rise of contemporary forms of consumption coincided with fears about the collapse of the family, the rise of the New Woman and the emergence of a scientific concern for 'sexual management', while Rachel Bowlby (1985) describes the 'dream world' of consumption as a 'seduction of women by men', metaphors that have been widely employed in the study of contemporary consumption.[2]

More recently, there has been much discussion of the apparent blurring of boundaries between shopping as 'work' and 'leisure', with the larger regional shopping centres describing themselves in the language of recreation and tourism. A typical example is the Merry Hill Centre outside Dudley in the West Midlands which invites customers to 'enjoy a wonderful day out' (Lowe 1993: 218). At its opening in 1976, Brent Cross made a number of similar claims. According to Peter Fenwick (of Fenwick's department store), centres such as Brent Cross had no need of 'side-shows' to draw in the public: 'We regard ourselves as part of the leisure scene. Shopping all day, family shopping is a leisure activity' (*New Society* 19 February 1976).

Many commentators would dispute this emphasis on shopping as leisure or entertainment, arguing that, for most women, shopping is hard work, a regular and routine job of domestic reproduction as well as a reinforcement of familial bonds. In the Canadian context, Prus and Dawson (1991) have examined the dual character of shopping as recreational and laborious activity, concluding that shopping is more likely to be defined as recreational if it is relatively free of social obligations and other pressures. Similarly, Colin Campbell (1993) argues that different kinds of shopping are viewed with different degrees of pleasure, according to the degree of 'autonomy' that consumers exercise over the activity: the more that 'choice' is involved, the more likely they are to enjoy the activity.

An exchange between two of our focus group participants makes a similar point. Debating the sexual division of labour in the home, Hannah suggests that men should do the cooking and cleaning while women do the shopping. Sean indignantly replies: 'I'd rather go shopping', to which Hannah retorts: 'That's the whole point, we ain't got a choice' (Wood Green Youth Project). While particular kinds of shopping such as buying gifts or Christmas shopping may be a 'labour of love' for some people, as Fischer and Arnold (1990) suggest, the work involved is rarely gender-free. As Rosemary Pringle (1983) reminds us, it is the association between gender and consumption that gives rise to its negative associations of extravagance and waste, contrasted with the positive (and masculine) associations of production as socially useful work. She cites Raymond Williams' (1976: 78) observation that, in almost all of its early English uses, *to consume* had an unfavourable connotation, meaning to destroy, to use up, to waste, to exhaust. Since the nineteenth century, consumption has come to stand for destructiveness, frivolity and insatiability, and shopping has been denigrated as a trivial and subordinated arena of 'women's work'.[3]

While we are keen to explore the relationship between consumption and identity, we are suspicious of arguments that downplay the extent to which the work of consumption is as active an exchange as the labour of production (cf. du Gay 1996). Although our focus group participants often employed the language of tourism in phrases like 'shopping trip' or 'day out' at the shops, our evidence suggests that the idea of shopping as recreation is currently overdrawn. Once the sexual division of labour and the gendered nature of shopping are recognised, the notion of shopping as leisure is much harder to sustain. Indeed, our focus groups included much evidence of the work of shopping including an emphasis on the skills required to be an effective shopper and the socially learned nature of 'good taste' (cf. Jackson and Holbrook 1995). As a member of the Wood Green Area Youth Project asserted: 'You have to learn to shop' – a view that was confirmed by several members of the St Andrew's Prampushers group in Brent Cross: 'It's something you have to learn', 'You have to learn it . . .', 'You do it because you've learnt it, it's a skill, isn't it?'

This was an issue in which differences emerged particularly clearly between the results of our three main methods. It would be quite possible to interpret the questionnaire results in such a manner as to affirm the current emphasis in the literature on shopping as a leisure activity. Respondents associate themselves with window shopping and browsing and with enjoying the experience. This remains a common element in the focus groups but it is tempered by an emphasis on both the skills and stress of shopping. In the ethnography, people continue to see leisure as an ideal of shopping and would like to see themselves as browsing and window shopping. But in day-to-day shopping this is matched by a very common feeling that people don't have enough time. Even many of the pensioners manage to construct their days in such a manner that they rival the housewife/parent by their aura of always being busy. One effect of this is that shopping is very rarely experienced as an unrestricted activity that can be undertaken as an act of either leisure or pleasure. In practice, informants generally shopped because they felt they needed to and with respect to particular items designated in advance. Any browsing or window shopping tended to become an interval or 'rest' from the requirement to shop for these required items.

The early history of Brent Cross provides ample evidence of the gendered nature of consumption and of its familial context. For example, late-night midweek shopping was originally introduced to attract families without disrupting men's traditional weekend activities:

> No longer will Dad have to make that near-impossible choice between watching his favourite football team or allowing Mum to choose the new double bed all on her own. As a mere male, I naturally take pleasure in predicting increased responsibility for fathers who will now be able to assist in decision taking in life, without having to forgo their hard-earned Saturday afternoon recreation.
>
> (Brent Cross Shopping Centre manager, interviewed in the *Hendon and Finchley Times*, special supplement on the opening of Brent Cross, 1976)

Other accounts of the opening of Brent Cross Shopping Centre place less emphasis on men's shared domestic responsibilities. An article in *New Society* (2 February 1978) described Brent Cross as an example of 'the woman's world of the second half of the twentieth century' where women shopped for pleasure while caring for their children in the scientifically approved manner of the times:

> As she pushes her pram up the main mall, the average shopper (25–34), moves as if being filmed for a building society advertisement. She talks to her offspring exactly as Dr Spock says she should.

More cynical views of 'Brent Cross mothers' were also expressed at around the same time:

> By lunchtime it gets more difficult to move. The play area is alive and aloud with small children, gratefully dumped by their parents. Brent Cross mothers have the reputation of forgetting their young. At anthropologically frightening intervals, a loudspeaker sends out pleas for mothers whose maternal instincts have deserted them.
>
> (*Observer Review* 1977 – exact date unknown)

Our survey data reveal that there were almost twice as many women (68.4 per cent) as men (31.6 per cent) at the two centres, with more women shoppers at Brent Cross (71.7 per cent) than at Wood Green (64.7 per cent).[4] This confirms the notion that shopping is a highly gendered activity, frequently regarded as 'women's work'.[5] More significantly, however, our survey evidence also reveals that relatively few people go shopping as a family unit. Sixty-five per cent of Brent Cross shoppers and 68 per cent of those at Wood Green reported that they shopped mostly on their own, with fewer than 5 per cent at either centre shopping with children or other family members (Holbrook and Jackson 1996a). Pedestrian counts (totalling 972 observations on Tuesday 12 September 1995) also reveal that family groups were a minority of those observed at the two centres (29 per cent of those observed at Brent Cross, 22 per cent at Wood Green).

People's stated shopping preferences are equally clear. Nearly 70 per cent of shoppers at both centres ranked shopping on their own as their first preference (see Table 5.1). Less than one per cent ranked shopping with other family members as their first preference, 2–3 per cent ranked shopping with children first and only 6–7 per cent gave top ranking to shopping with their spouse.[6]

While comparatively few people actually go shopping as a family (and even fewer prefer to shop in this way), we argue that 'family shopping' as a social construction has relevance for almost every consumer including those who live and shop alone. Even single-person households encounter representations of the family wherever they shop. As we shall see, however, constructions of the family are frequently expressed in relation to other discourses such as those related to 'race', crime and neighbourhood change. In order to explore the nature of 'family shopping', we concentrate first on one or two key relationships and then broaden the discussion to include a series of related discourses.

Table 5.1 Stated shopping preferences

Brent Cross shopping centre *(percentage)*	*Rank 1*	*Rank 2*	*Rank 3*	*Rank 4 or lower*	*No answer*
On own	67.9	14.3	7.0	4.7	6.0
With a friend	10.5	44.4	11.4	5.4	28.3
With group of friends	7.3	6.3	9.5	2.9	74.0
With spouse	7.3	8.6	9.2	2.2	72.7
With parent	2.9	3.8	11.1	9.2	73.0
With child	2.5	7.6	7.9	2.8	79.0
With boy/girlfriend	1.6	7.6	6.0	5.1	79.7
With family	0.3	1.6	3.5	0.6	93.9
Wood Green Shopping City *(percentage)*	*Rank 1*	*Rank 2*	*Rank 3*	*Rank 4 or lower*	*No answer*
On own	65.7	15.0	8.4	3.0	7.7
With a friend	15.7	40.6	14.3	2.1	27.3
With group of friends	4.9	8.4	9.8	5.2	71.7
With spouse	5.6	8.7	8.0	3.4	74.1
With parent	2.5	4.3	5.3	11.6	76.2
With child	3.1	9.1	7.7	3.8	76.2
With boy/girlfriend	2.8	5.6	8.4	7.0	76.2
With family	1.0	0.7	0.7	1.6	95.8

Note: Respondents were asked to rank each of the categories (on own, with a friend etc.), with 1 indicating their strongest preference. 'No answer' indicates categories which were not given a ranking.

Source: Survey results

Mothers and children

People's reservations about shopping in family groups are not hard to fathom when one talks to young mothers about the problems of shopping with small children. Several of the group discussions focused on these problems. Mothers complained bitterly about the lack of adequate toilet facilities at both centres and about escalators that were impossible to negotiate with bags of shopping or baby buggies (pushchairs). For women in the Devon Close Mother and Toddler group, for example, there was no pleasure in shopping with small children: 'With two kids in tow, it's no enjoyment.' Similar views were expressed about 'dragging the children shopping' to Brent Cross by women in the Finchley and Whetstone National Women's Register group:

> some of my negativity towards Brent Cross goes back to the days when the children were small and I was trying to struggle with buggies and things in Brent Cross because it's like all shopping centres, it's not really designed for mothers with children . . . The children's departments or baby stuff is always at the back of the shops so you have to struggle all the way through – it's not 'parent friendly'.

Other members of the same group agreed: 'It's stressful with children . . . there's always a huge queue; it's very difficult with young children getting around.' A member of the St Andrew's Prampushers group encapsulated many other people's views:

> I hate Brent Cross . . . It's too expensive, it's too hot, there's not enough security guards around, there's not enough toilets on every level, so you've got to queue up for two hours with a two-year-old who can't hold on.

Both centres have tried to address these concerns by providing feeding, changing and toilet facilities which are, in fact, much better than the facilities provided in most conventional high street shops. Furthermore, when seen comparatively, the enclosed spaces of the shopping centres were found during the ethnography to be often preferred to the high street, since there was protection from excessive sun or rain as well as the possibility of allowing children out of the buggy without the danger of their running into the road. The fact that mothers expressed such vehement views about their inadequacy may represent a displacement of other feelings they have about shopping with their children onto the centres themselves. For 'family shopping' certainly generates strong feelings of resentment and guilt, deferred pleasure and personal denial.

In particular, our evidence suggests that parents' identities are often constituted through emotional and financial investment in their children and that such investment may be experienced as a loss of personal autonomy, expressed through nostalgic reflection on the pleasures of shopping before children were born. For example, one of the participants in the National Women's Register group claimed that she only went to Brent Cross 'for my children's sake'. Another concurred: 'I am drawn there by my children very often because they like the range of sports shops and so on. Other than that, I avoid it, [going] only when necessary.' Another mother in this group compared her own sense of shopping as drudgery with her teenage daughter's obvious pleasure at shopping with her friends:

> My daughter's thirteen and her idea of an outing on Saturday is to go on the bus with one of her friends, and they don't spend very much but they look at everything in the Body Shop and Miss Selfridge and Top Shop and Gap and all those places, and it's fun. Nightmare [to me], but to them it's fun.[7]

Others agreed. Young people go to Brent Cross for the whole day ('they parade, it's their playground'), while for mothers with young children the experience is very different ('you want to get in and out without them whining and going on all the time') (all National Women's Register).

By contrast, shopping with friends can still be a source of pleasure for these women, worth 'dressing up' for and 'parading around':

> I had a girlfriend from Bristol staying with me once and we just . . . had to go over there *[Brent Cross]* and I just, as I am now, just jumped in the car and went. And we got there and she was really cross with me because she thought everybody was dressed up and . . . she said 'You didn't tell me everybody got dressed up' and I thought 'Oh I didn't, I don't look at people', I mean I just didn't *[laughs]*. 'Cause she was really upset, she would have liked to have paraded there for hours.
>
> (National Women's Register).

This group was a relatively affluent one, but all the women remembered the time before they had children as a period when they had more money to spend on themselves, when shopping was more fun:

> I used to like Brent Cross. I remember when it first opened. I enjoyed shopping there. But then my life was very different. I was single. I had no kids . . . I had money to spend.
>
> (National Women's Register)

The situation is even more marked in working-class groups such as the Devon Close Mother and Toddler group in Wood Green. As Deirdre observed:

> It's alright when you're going to get something that they want *[her children]*. It's when you want to look at something that they get fed up.

Asked to think back to the time before they had children, all the members of this group agreed:

> Ooh, I loved shopping.
>
> I loved it.
>
> There was so much money to spend.
>
> It was the bestest thing.
>
> It was the most relaxing thing to do.
>
> We'd go out every weekend and spend all the money.
>
> It didn't matter, you didn't have to worry about what the kids was gonna have next week. When you're on your own that's best.
>
> As soon as you have kids, it changes all of that 'cos you've always got to worry are they going to need something by the end of next week. Are you spending money you're going to need by the end of next week?
>
> (Devon Close Mother and Toddler Group)

Before they had children, there was more time and money to spend on themselves, circumstances that are remembered with a kind of defiance:

> I love shopping, love buying things. The only thing I don't like is, I won't go unless I have the money. If I've got the money, yeah, I could shop, I could shop from now 'til next week. If you give me the money, it doesn't matter how many kids I've got, I'll let them know, I love shopping.
>
> (Devon Close Mother and Toddler Group)

Many of our respondents reflected with almost guilty pleasure on those more carefree days: 'I spent an awful lot when I was younger', 'You had more time to spend on yourself', 'when you haven't got anybody else to spend your money on, you spend it on yourself' (all National Women's Register). Several (mostly middle-class) mothers spoke of the guilt they felt, shopping for themselves:

> I feel guilty spending money on clothes and things like that, 'cause they're so blinking expensive ... I don't often splash out on luxuries, as I said before, I feel guilty. But you know, family life, you do just spend money all the time. But it's necessity shopping rather than luxury shopping.
>
> (National Women's Register)

This distinction between 'luxury' goods and shopping for 'necessities' seems to be more to do with who the goods are for as simply a matter of price. Women can be seen to be defining themselves as mothers through the sacrifices they make to their families, particularly their children. As Deirdre of the Devon Close Mother and Toddler Group said:

> Don't you find, if you go shopping, this is my problem always, I go shopping for myself, and I say, right this is it, I am going out today, I am going to get something for myself and I will come back, guaranteed, with something for everyone else at home except what I want . . . It's like I get money for Christmas or birthdays or whatever, and I decide, right, that's it, I'm going out and spending this on me and I go out and I see something and I think, oh yeah, that would be nice for Kylie *[her daughter]* and then I think I can't get something for Kylie without getting Paul something *[Kylie's brother]*, and then I think, I've got something for them now I've got to bring something home for Mick *[Deirdre's husband]* . . . I come back and I'm lucky if I get, like, say an eyebrow pencil or an eye-liner out of it. I intended to go out and spend maybe £30 or £40 on myself and I go out and spend on everybody else and I'm lucky if I get two quid out of it.

Personal shopping for clothes or other 'luxuries' provoked a sense of guilt rather than pleasure as these mothers surrendered their sense of self to the needs and wishes of their families: 'I regard shopping for myself as a luxury', 'I have to go really for my children's sake'.

These mothers clearly defined shopping as their responsibility: 'It's a woman's job isn't it really.' They were either scathing about their partners' lack of skill or interest in shopping ('I don't know any man who likes shopping', 'My husband's taste I can't stand', 'If my husband goes shopping, the bill is £20 more') or excused them on the grounds that they didn't have sufficient time ('he doesn't have much time really for shopping', 'they go out to work and their relaxation time is at home and they don't *need* to go out and look round the shops', 'they haven't got as much spare time as us' – all Devon Close Mother and Toddler Group). Central to this sense of responsibility was the need to take proper care of their children. According to Deirdre: 'Personally, I wouldn't let either of them out of my sight . . . She's only got to run out of my sight and I leave all the shopping to go and get her, then I've got to leave the other one'.

Some mothers relied on other family members to provide childcare while they shopped:

> Whenever I go to Wood Green, I always leave me children either with me Mum or with me husband. I never take them to Wood Green. I'll maybe take the two-year-old but, obviously, I can't take both of them on the bus and you can't concentrate with two, it's difficult.

Very few were prepared to leave their children in a crèche or other facility where childcare was provided by strangers:

> I couldn't. The only way I'd be happy with leaving the kids is if it's some-body I know, 'cause there's no way I could walk into a crèche in a shopping centre and say, 'There's my kids, I'll be back in an hour'. You don't know that person from like, you know, Jack the Ripper.

In both practical and emotional terms, for mothers with young children shopping involves a complex balancing act between conflicting interests and responsibilities (cf. DeVault 1991). One member of the St Andrew's Prampushers told us that she had waited two years to buy some new trainers. She had been given money for Christmas but, as she put it, 'spent it on that little squirt'. Other members of the group explained that their children 'always need something, don't they' and 'you like to see them well dressed up'. While parents clearly derived great pleasure spending money on their children, investing emotionally and financially in their well-being and appear-ance, it was also mildly resented as a loss of their own autonomy and independence.

There were several instances where mothers complained of spending less on themselves and more on their children: 'If I'm given money [for Christmas], I spend it on the kids' (Devon Close Mother and Toddler Club). Sometimes a sense of resentment crept through. As Karen from St Andrew's Prampushers said:

> I'm getting selfish now. I'm not buying him *[her son]* so much. I'm getting my Mum and my mother-in-law to buy his clothes so that I only have to buy socks and underwear, and now I've got money to decorate. Not that my husband will do it, but I've got the money to do it with.

Despite feeling 'selfish', even in this instance Karen was contemplating spending money on the family home and not on personal items for herself. Another member of the same group described Oxford Street as 'out of bounds' because of the temptation of personal shopping there, while another talked about the guilt she felt when putting her own wishes ahead of her children's needs:

> I'd like to be able to go into a shop and buy something for myself without feeling guilty about it . . . We'd like to go and treat ourselves to some-thing in the shops but you can't because your children always want something.

If parents' identities are constituted through relations with their children, we found that teenagers commonly used shopping to express their growing (and contested) sense of autonomy from their parents.

Teenagers and parents

Some of our teenage respondents were aware of the 'sacrifices' that their mothers made on their behalf. As Stacey, from the Wood Green Area Youth Project, observed:

> My Mum ... don't think of herself. She thinks of us two first. It's like, if we go shopping, she'll think of us first and she'll only get herself something every two months, like, get herself a really nice outfit. Like, kids are first, you get what I mean?

Another member of the same group expressed a similar view:

> I think my Mum always buys me and my sister stuff and like, every now and again, she goes to buy herself something. So when she goes out and buys herself something, [if] she spends a lot of money I feel good as well, as I think she deserves it ... My Mum works so hard, just by bringing us up, so I think she deserves a good time and to go out and enjoy herself.

But for the majority of teenagers, shopping with their parents was regarded as a source of tension, best avoided if possible.

Several of our teenage respondents used shopping as a way of marking out their growing independence from their parents. One member of the Greek Cypriot Youth Group in Wood Green summed up many young people's views: 'I feel embarrassed being seen with my Mum.' Lorraine, from the Wood Green Area Youth Project agreed: 'I don't like shopping with my Mum and Dad 'cos my Mum always chooses the clothes for me and, if I don't like it, it's tough.' Alex had a similar view of food shopping with his mother: 'I don't really choose things. My Mum sits there picking things up off the shelf and I'm just chasing her with the trolley.' Participants at the Canada Villas youth group recalled frequent disagreements with their parents:

> If you've got your Mum with you or your Dad, it's like, 'No I don't like that.'

> And they start going, 'Oh what's taking you so long, don't you want this and don't you want this?' You don't get a chance.

> That's too short, that's too long, you're showing too much of your legs.

> That's obscene!

> No, you can't wear that.

As this passage suggests, teenagers often had different criteria from their parents, placing much more emphasis, for example, on designer labels in clothing, jewellery and particularly sports shoes: 'Sometimes it takes me up to three days just to choose a pair of trainers', 'Yeah, it's gotta have "Nike" ... if you're wearing a pair of Dunlop they'll laugh in your face' (both Canada Villas). Teenagers were extremely conscious of style and fashion, drawing an explicit connection between their 'identity' and what they wear (and how they wear it):

> Yeah, you want to be in the fashion, don't you. It gives you your own identity. 'Cause if you've got your own, like if you go into a shop, it don't matter if hundreds of people have got that same piece of clothing, it's how you wear it and what you wear with it ... We buy what's in fashion, really, 'cause you don't wanna be wearing a shell suit when everyone else is in like.

> *[ridiculing]* Shell suit?

> You buy what you like, you don't buy what someone else has got because it's in fashion. You buy what you want, don't you. You buy what you like ... if I know it looks alright on them, I'll go and buy one.

For younger teenagers, parents still tended to control their expenditure, causing further frustration. Laura complained about her mother's taste which differed markedly from her own:

> My Mum will say ... like I want a pair of jeans for £25 and they're good quality. My Mum will go for £15 and tell me to get that. My Mum will tell me, you go for that if you want to save up your own money. I'll buy you this, but I won't buy you that. That's what I hate. Say, like you're a teenager now ... and you want to choose your own stuff. My Mum doesn't give me a chance to do that. That's why I hate shopping with my mother.
>
> (Wood Green Youth Project)

Similar issues arose for Zoe, one of the participants in the Greek Cypriot youth group:

> We have arguments over shoes, you know. I've always had. I really like rock music and I like the rock look, you know. I like DMs *[Doctor Marten's]* so my Dad doesn't allow me to wear them. This year, he gave me money and said, 'Your Mum is going with you to buy some shoes' *[she laughs]*. I thought, I'm 18 years of age, I can do it myself. ... Yeah, he made me take a pair of boots back just because they were up to here *[pointing to mid-calf]*. You know, he's a very strict man ... I bought a football top once and he had a go at me and started shouting and everything. So it's really, you've got to ask him what you can wear and what you can buy.

103

Zoe clearly resents the influence her parents' continue to exert over what she can buy but has little choice because she is still financially dependent on them.

Several parents felt that their children had 'grown out' of Brent Cross: that it was an appropriate place to shop for children's and adults' clothes, but had less to offer teenagers: 'My sons are 16 and 20 and they wouldn't be seen dead in anything from Brent Cross now' (National Women's Register). Participants at the Canada Villas youth group agreed that Brent Cross tended to 'skip our generation . . . It's like they've missed out a whole generation.' Arguing that 'men have got more style than they used to', James preferred to shop 'up London' in Camden Town and Tottenham Court Road where it's 'more trendy, more fashion', 'special names, drug clothes like Spliffy and Clinton'. Significantly, the current extension and refurbishment of Brent Cross is designed to help target this younger clientele, though still emphasising the 'family environment' that most shopping centres aim to convey.

A family environment?

Most shopping centres, from the most spectacular 'mega malls' to more mundane shopping centres such as Brent Cross and Wood Green, present themselves as 'family environments', suitable for people of all ages and particularly welcoming to parents with small children. They are consistently presented in this way, though, as we have seen, they may not be perceived as particularly parent-friendly. For example, the Meadowhall centre, outside Sheffield, describes itself as 'an ideal destination for all the family' (Meadowhall Mall Guide). Claiming to be the most parent-friendly centre in the UK, Meadowhall provides parent and baby rooms with baby food and bottle-warming facilities, high chairs at all the restaurants in the 'Coca-Cola Oasis' and a crèche facility (sponsored by Typhoo Tea) with a money-back guarantee if your child is not settled down within a few minutes. Similarly, the Merry Hill centre in the West Midlands claims to provide 'Safe, secure family shopping at its best', with a children's play area and free mall entertainment (Lowe 1993).

Managers at Brent Cross and Wood Green also saw 'family shopping' as a key element of what they provided for their customers. During interviews, both managers claimed to provide a 'family environment', emphasising 'family values' and catering to the needs of everyone, but above all to families. This seems to run counter to the evidence presented above, that relatively few people actually shop as a family unit and that those who do so rarely enjoy the experience. We suggest that this apparent contradiction points to the metaphorical function of 'the family' as a marketing strategy. References to shopping centres as 'family environments' serve as a kind of shorthand for the kind of ambience that shopping centre managers are trying to create (clean, warm, safe, friendly, welcoming). As one US shopping centre consultant argues, to maintain the 'family appeal' of shopping malls: 'You have to create a safe, secure feeling and make sure it's not intimidating to anyone' (quoted in Goss 1993: 28). An emphasis on 'family values' serves to create an appropriate place-image even if, in practice, relatively few people actually shop as families. Despite general agreement on the 'crisis' of the contemporary family (McRobbie 1994), the word still seems to act as a powerful and comforting metaphor for social cohesion and stability, ideas that are constantly undermined by people's fear of crime and insecurity in many urban areas (cf. Beck and Willis 1995).

In many of our focus group discussions, we found that the pleasures of shopping were tempered by widespread fears, focused on particular places and social groups. These discussions lead us to conclude that the popular appeal of shopping centres such as Brent Cross and Wood Green derives in no small part from their ability to 'manage diversity', using the power of privatised space (including surveillance and other more informal means of social control) to exclude those who are considered 'undesirable'. By managing the mix of retail outlets and controlling accessibility to the centres (if necessary by forcible ejections and exclusions), the managers of these spaces are able to reduce the risks of unanticipated encounters with people they would find threatening, while actively promoting the virtues of 'familiarity'. These strategies can be likened to a process of 'domestication', making shopping safer for the majority of middle-class consumers at the cost of social and spatial exclusion for others (mainly young, racialised, working-class men).[8]

Fear of crime

Both Brent Cross and Wood Green shopping centres are highly conscious of the impact of crime and insecurity on their attractiveness to consumers. In the recent past, both centres have experienced major security problems involving terrorist bombings (see chapter 2). In 1991 there was a major bomb blast at Brent Cross on the last Saturday before Christmas ('Fire bomb blitz', *Hendon and Finchley Times* 19 December 1991). Over 100,000 shoppers were locked out of the centre and lost revenues were estimated at several million pounds. Four bombs were discovered of which two exploded in C&A and Fenwick's. In December 1992, a bomb exploded outside W H Smith and the Argos catalogue store near Wood Green Shopping City, causing minor injuries to ten people including four police officers ('Police and shoppers injured in bomb blasts', *Independent* 11 December 1992). As the manager remarked, although events such as these may have taken place in other parts of the High Street, many people associate them with the Shopping City itself.

Significantly, however, it was the risk of street crime rather than terrorist violence that provoked most widespread fear and it was here that our evidence suggests most difference between the two centres. Judging by our survey results and focus group discussions (as well as an extensive review of local newspapers), Wood Green had a worse reputation for street crime and burglary than Brent Cross. While some focus group participants complained that Brent

Cross 'attracts a very big criminal element' (Brent Cross Residents' Association), others found the level of security there very reassuring:

> compared with an open air shopping parade, I would think I'd feel much happier at Brent Cross 'cos it's lit and there are security guards, there's always plenty of people.
>
> (Unitarian Church group)

Brent Cross has a reputation for motor vehicle crimes including car theft, but this gave rise to far fewer concerns about public safety than the perceived level of street crime in Wood Green. Whereas 42 per cent of survey respondents in Brent Cross described the centre as 'safe', only 25 per cent did so in Wood Green (see Table 3.8, p. 60). Most respondents' anxieties centred on the fear of robbery and inter-personal violence (mugging). As an Indian woman in the English as a Second Language job club told us: 'I don't like to walk at nights in Wood Green Shopping City or anywhere because I'm scared . . . I always go by car.'

The local press (whose significance in the dissemination of fear of crime is discussed in Smith 1985) is full of stories about crimes of violence in the Wood Green area: 'Beast boy faces a life sentence' (*Haringey and Wood Green Independent* 23 September 1994); 'Cashpoint man acts to beat off muggers' (*Haringey and Wood Green Independent* 23 September 1994); 'Thief ko's shop boss in £13,000 raid' (*Haringey and Wood Green Independent* 30 September 1994); 'Mad arsonist strikes twice' (*Haringey Advertiser* 13 July 1994); 'Armed burglars raid home' (*Haringey and Wood Green Independent* 7 October 1994); 'Gang kicks victim to ground' (*Haringey and Wood Green Independent* 8 July 1994). Wood Green Shopping City was also identified in the national press as the location of a recent stabbing and a place where organised shoplifting takes place ('Child shoplifters work to order', *Independent* 19 January 1994). On average, crimes were reported about seven times a month around Wood Green Shopping City in 1993 according to the Town Centre manager (press release, January 1994).

Significantly, the centres have responded very differently to their threatened security, revealing a sensitivity to their contrasting customer base. At Brent Cross, with a high proportion of White, middle-class and Jewish customers, the response has been high-profile. Security guards in luminous jackets patrol the car parks. Surveillance cameras are conspicuous and indoor guards patrol in uniforms that resemble those of police officers. At Wood Green, by contrast, with a higher proportion of black and working-class customers, a much more low-key approach is evident. Surveillance cameras are much less visible and security staff are dressed in maroon jackets and dark grey trousers to distinguish them clearly from police officers. Despite the shared problem of insecurity, the managers at both centres are well aware of these social differences, reflecting their local knowledge and sensitivity to their differing clientele.

While most respondents at both centres welcomed the introduction of closed-circuit television and other forms of customer surveillance (and there are plans to extend these to the whole High Street), there were some notable exceptions, particularly among ethnic minority teenagers, who felt they were the target of such methods.[9] Focus group participants in the Wood Green Greek Cypriot Youth Club expressed a common anxiety:

> I was followed by two security people just because of the clothes I was wearing. If I went in like this *[smartly dressed]* after work in a suit or whatever, they wouldn't even give me a second look. But I was young, I was in a tracksuit . . . and automatically you know it's 'shoplifter'.

Another member of the same group agreed:

> It scares me, all this security . . . It's like they're trying to pry into all aspects of your life. You can't even go two miles over the speed limit and you're going to get booked. It's going to become like 1984, George Orwell, you know, like Big Brother and everyone spying on you. Next we're going to have a camera in the corner over there to listen, just in case we say anything against them . . . it's terrible.

Members of the Wood Green Area Youth Project had had similar experiences:

> I walked into W H Smith and the security guard followed me the whole way round the music store.

> It's intimidating 'cos they're watching you; it makes you feel a bit dodgy.

A young woman in this group reported her own experience:

> Before Christmas, yeah, the security guard comes up to us. We were standing there, we weren't doing nothing wrong, weren't causing no trouble. We wasn't doing anything to nobody and the security guard says to us 'Can you move along, please?'

Participants in the Canada Villas Youth Club group had also been watched and followed round at Brent Cross:

> you get kicked out *[by the security guards]* for doing nothing, for doing nothing whatsoever, just standing there looking over a balcony. All of a sudden, these guards start crowding round us and chasing us round Brent Cross and they kicked us out.

There is also some evidence to suggest that respondents' fear of crime may have been exaggerated in relation to declining official levels of reported crime.

According to police statistics, Haringey (where Wood Green is located) is by no means the most crime-ridden London borough. The figures for Barnet (where Brent Cross is located) are even lower. Yet, as has been noted elsewhere (Smith 1987), fear of crime is highly localised.

Despite local people's concerns, the level of reported crime in Haringey actually went down by 6 per cent between 1991 and 1992 according to police statistics (*Haringey People* November 1993). Following a series of high-profile policing campaigns, the rate of burglary in Hornsey also went down, with motor vehicle crimes now 'almost off the scale' (Wood Green Open Forum, 4 September 1995). Fifty-seven arrests had been made over the previous three months on Wood Green High Street and the clear-up rate had increased from 3 per cent to over 15 per cent. As the Chair of the Open Forum noted, 'Crime in car parks is way down but there is still a high perception of crime which is a problem in attracting people to the area'. Understandably, the Wood Green Town Centre Manager was keen to publicise this decline in reported crime.

Levels of reported crime in Barnet are much lower than in Haringey. The highest figures are for car crimes with 5,648 offences reported in 1993. Figures for violent crime are much lower and clear-up rates are improving (*Barnet Profile*, London Borough of Barnet 1994). In our focus group discussions, Brent Cross shoppers were keen to dissociate themselves from the perceived incivilities of Wood Green: 'Wood Green isn't part of this area', 'It's a bit scary, Wood Green', 'more violent', 'some of the teenagers round here are a little frightened of Wood Green' (various members of the National Women's Register group).

The discrepancy between (declining) levels of reported crime and (rising) levels of insecurity suggests that 'fear of crime' may be acting as a vehicle for other issues. All the evidence reviewed by Smith (1987) suggests that fear of crime differs amongst social groups and between urban neighbourhoods. Fear of crime, she suggests, is particularly debilitating for women and the elderly. Furthermore, as evidence from the United States suggests (Taub et al. 1984), fear of crime is often associated with people's attitudes towards the changing racial composition of urban neighbourhoods. Both of these themes are borne out by our evidence from Wood Green and Brent Cross where the issues were often expressed through a sense of nostalgia for a previous 'golden age' of shopping that had been lost through the 'retail revolution' of the last few decades (cf. Brewer and Porter 1993; Wrigley and Lowe 1996). The ethnographic study also suggested that nostalgia was particularly associated with the elderly and that pensioners had the most extreme fear of crime, often refusing to leave their home once it became dark.

Nostalgia and the racialisation of neighbourhood fear

In some cases, people's attitudes to shopping in the past exemplified a generalised nostalgia ('It used to be a joy to shop'). In other cases, it took a much more specific form. Members of the Brent Cross Residents' Association, for example, felt that a sense of local particularity had been replaced by increasing standardisation:

> Shopping is not what it was, I mean shopping used to mean going to the corner shop or going to the butchers and doing a specific ... but shopping now, it's totally different because 98 per cent of all households today own a freezer.

Similar arguments about standardisation and lack of choice came across in group discussions at the Woodside Women's Group:

> There's been all these big take-overs, they bought so many shops ... they've all got the same sort of thing, haven't they?

> Once you've been to one shop, you've been to them all.

> You get shoe shops, they've got everything the same – clothes, they've all got the same thing, not like years ago when all the different shops had everything different, didn't they?

Expressions of nostalgia often overlapped with some of the other discourses that we shall discuss in due course (particularly in chapter 6), concerning the decline of personal service and the increasing 'artificiality' of the built environment. Here, however, we focus on people's fear of change at the neighbourhood level and, in particular, the racialisation of such fears.[10] Many of the older participants in our focus groups were unsettled by the global sourcing of fresh produce: 'you get stuff from all over the world these days' and 'you don't know how they're grown' (Unitarian Church). Participants in the Wood Green Bowling Club group made a series of interesting connections between increased standardisation and the changing character of the neighbourhood:

> There are too many shops that are the same.

> I don't think there's enough variation.

> It's all one thing now ... there's too much of the same thing.

> ... it used to be beautiful, lovely shops. Wood Green used to be called the Golden Mile.

> When there were ordinary shops and that sort of thing down there, the shopkeepers got to know you and you got to know them.

> If there's any corner shops left, either the Indians, Pakis or what have you, sweep it up. You see ... Haringey is completely spoilt by ... they admit

to 80 per cent ethnics in Wood Green and Haringey. Where we've worked all our lives and now we're strangers because they've taken over.

(Woodside Bowling Club)

For our focus group participants in both areas, but especially in Wood Green, feelings of nostalgia were often articulated through the language of dirt and pollution: 'The shopping in the area has changed, there's a sort of deprivation creeping into the local shops . . . they sell stuff dirt cheap' (Brent Cross Residents Association). The 'old values' of cleanliness and quality are contrasted with the 'rubbish' on sale today, as in this exchange among members of the Woodside group:

> There used to be a linen place as you went down Wood Green. What was that called, Hopkins?
>
> Hawkins, yes, for linen, yes, real linen, it used to be beautiful.
>
> Nice quality shops used to be in Wood Green High Road.
>
> Yeah, it's rubbish today . . . Taiwan, Hong Kong, Jamaica.

At first sight, this might be interpreted as a simple preference for department stores and specialist outlets (with their associated benefits of 'personal service') over the perceived anonymity of shopping malls and supermarkets. But the desirability of particular commodities from 'nice quality' local shops ('real linen . . . beautiful') is clearly contrasted with foreign 'rubbish'. The connection is repeated later in the same discussion, this time in relation to inferior imported goods ('foreign muck'):

> . . . it was good years ago, but it's not any more.
>
> It's still all foreign muck, isn't it – that's the way I look at it anyway. I mean look at the trouble I've had with cameras . . . Hong Kong muck.

The same theme recurs in another exchange about the lack of personal contact associated with modern shopping centres:

> *[in the old days]* . . . you'd meet people and say hello . . . everybody knew one another, but you don't now.
>
> I can go over Wood Green and not see a soul I know.
>
> That's right. I've been out all day and not spoken to a soul.
>
> But it's altered, hasn't it? I mean England, hasn't it. It's all altered.
>
> It's multi-racial isn't it, after all, you get all kinds.

(Woodside Bowling Club)

Concerns about dirt and pollution were also commonly expressed via an obsession with the condition of public lavatories.[11] Compared to the 'spotlessly clean' toilets at Edmonton Green, those at Wood Green Shopping City and at Turnpike Lane (the other end of Wood Green High Road) were described as 'filthy': 'the ladies' toilets, sometimes you go there and there's no paper anywhere . . . it's nearly always half cut off, you know, something's wrong, it's bad' – all Woodside Women's Group). This was contrasted with an idealised past, when 'road sweepers [were always] clearing, sweeping through the High Road' ('They were spotless on a Sunday morning, weren't they?'), culminating in an exaggerated celebration of the virtues of the past: 'I mean, when we was children, they could play marbles in the gutter. Well, who could do that nowadays?' (all Woodside Women's Group).

The language of dirt and pollution has long been associated with discourses of 'race' as Anne McClintock's (1995) work on commodity racism and imperialist advertising amply demonstrates. Victorian fears for various racialised Others were translated into domestic concerns for cleanliness and purity. Advertisements for Pear's soap, for example, imbued the product with the magical property of turning dark skin white, while Eno's liver salts were revered throughout the British Empire for their ability to cleanse the 'inner man'. In contemporary London, a nostalgia for the past is frequently expressed in similar terms where the purity of 'real linen' or 'nice quality shops' is contrasted with 'dirt cheap' prices for 'imported rubbish' and other 'foreign muck'. Similar issues were sometimes expressed through an exaggerated respect for the cleanliness of long-established stores, like Marks & Spencer, now considered archetypally 'English' despite its immigrant Jewish roots:

> As standards go, you can't get much higher than Marks & Spencer's. I mean, you know, they carry very, very high standards – more than people realise – they are so meticulously clean, it's incredible.
>
> (Brent Cross Residents Association)

The racialisation of neighbourhood change was particularly clearly expressed in relation to food and eating rituals where 'traditional' family meals and 'home cooking', including a 'proper' roast Sunday lunch, were contrasted with the self-indulgent and extravagant move towards snacking and fast food:

> A lot of the young ones, they don't know how to cook a proper meal like we used to, which works out our way a lot cheaper. You could make a big meal for a family couldn't you?
>
> I used to buy a chicken. I used to cook it. There was three of us. I used to cook it and have it one night and then perhaps cut it up the next, then I used to boil the bones, trim it all off and make pasties, that lasted me nearly a week. But they can't do that now.
>
> (Woodside Bowling Club)

Again, in this exchange, the conversation quickly moved on to questions of 'race':

> Next door neighbours . . . they were marvellous.

> You could leave your key on a bit of string through the letter box for the neighbour to come in, never knock at the door, just pull the string through, open the door and shout out, 'It's only me Brownie', that sort of thing.

> You don't know your next door neighbours, cause there's so many foreigners about, isn't there . . . I mean I got nothing against them at all, but when you get them all mixed up as foreigners, it's not fair.

> And they haven't put a penny into the system, this is what annoys you. We've paid out all our lives, I mean from when we're 14.

> I mean, they're quiet. I've got nothing against the people at all, only their stinking cooking [*general laughter*].

These elderly White residents cheerfully contrasted the 'stinking cooking' of foreigners to their own preference for 'old-fashioned fish and chips . . . roast beef and all that' (all Woodside Bowling Club). Their endorsement of all things English reached its apogee in their affection for 'a really good cup of tea', blissfully unaware that tea-drinking only became popular in this country in the nineteenth century and that it was reliant on a far-flung network of imperial connections (a point that is nicely elaborated by Cook and Crang 1996).

Conclusion

In this chapter we have argued that 'family shopping' is one of the key contexts of contemporary consumption in the sense that most shopping decisions are made with respect to a relatively limited range of social relations. Our evidence suggests that, for centres such as Brent Cross and Wood Green, the domestic sphere of 'family shopping' is far more prevalent than the hedonistic pleasures of 'lifestyle shopping' described elsewhere in the literature. We also suggest that the popularity of shopping centres and malls is related to the perceived incivility of more public places such as the high street where random encounters with racialised Others are much less easy to control. Shopping centres work by managing such diversity, providing a 'domesticated' shopping space where middle-class consumers feel safe, though this is achieved by the surveillance and exclusion of those who do not have the appropriate degree of familiarity. The managers of such places market their centres as places for 'family shopping' even though the majority of their customers rarely go shopping as a family unit (and those who do would prefer to be shopping alone or with friends).

This point has been extended by an examination of the racialised fears of elderly White residents in Wood Green, who have experienced a loss of familiarity and control over their local shopping centre. They express these feelings through a language of dirt and pollution, closely associated with the changing 'racial' composition of the neighbourhood. Their reflections on the level of crime in the neighbourhoods also take on a distinctly 'racial' cast, even where external evidence suggests that a reduction in the actual level of reported crime may be occurring. The chapter reveals how social identities are constituted through boundaries of inclusion and exclusion, articulated in terms of family relationships and by distancing those who are perceived to be socially different. Shopping centres like Brent Cross and Wood Green have come to represent an idealised version of high street shopping where the fear of Others has been thoroughly 'domesticated' through an emphasis on familiarity and a general absence of unwelcome social difference. While high street shopping may be recalled nostalgically as more spontaneous and less artificial (with more emphasis on social contact and 'personal service'), people's actual shopping practices now clearly favour the planned environment of shopping centres like Brent Cross and Wood Green however sanitised the experience is said to have become.

6

THE NATURE OF SHOPPING

Introduction: naturalising shopping centres

During the analysis of our field material one of the authors (Miller) noted that as someone who grew up with Brent Cross, who remembers its original opening and for whom it has always been a major shopping location, the recent changes to that centre have come as something of a shock. Of the many people whose comments he could recall from casual encounters over the years, no one ever had a good word to say about the aesthetics of the Brent Cross exterior, the way it presents itself to the outside world, in particular, the traffic on the North Circular. On the other hand it was rare that a bad word was said about the Brent Cross interior, and the high central dome had become an especially favoured and familiar site to regular shoppers. The fountain below and the mock stained glass above had been readily assimilated and seemed to work extremely well as the key point of identification that made Brent Cross special. If the site had not had such a sensitivity to commercial immediacy he would have predicted that this would have remained an ever more familiar and in a sense friendly part of urban popular architecture, destined to be recognised and protected by some national heritage committee as a key memorial of the 1970s. Indeed on some late afternoons, as the light gradually faded and the colours of the mock stained glass in the dome themselves started to homogenise to a universal dullness one could easily evoke similar experiences in the great religious monuments visited as a tourist and perhaps the same slight twinge of spirituality might have been evoked.

During the course of this study, however, first the mystique and then the actual structure itself was demolished. One of the first events in the initiation of our study was a tour behind the scenes, arranged for us as a group of researchers by the manager of Brent Cross. During this tour we went above the dome to see the actual fibreglass panels from which it was constructed and to remove a stained plastic panel to gaze on the shoppers below. At the same time we heard that the decision had been made to replace this and the other roofing in order to open out Brent Cross to natural light, at a cost which is hard to comprehend for those who live only with domestic budgets rather than commercial concerns.

A few days later we were discussing the changes in Wood Green with the manager of that site to be told once again that vast sums had also been spent a few years previously on opening up the site to natural light. Clearly these changes have not occurred in isolation. On 5 August 1995 the *Estates Gazette* reported a statement from the owners of Brent Cross, Hammerson, about the transformation which would add 4,646 square metres (50,000 square feet) to the centre: 'for us therefore, the starting point has to be thorough research ... our objective is to ensure that the duration of shopping trips is extended. Our research has shown that the longer the trip the higher the expenditure!' The caption to the accompanying photograph says that 'light is a key factor in new designs for the centre'. One implication seems to be that opening up to 'natural light' might somehow be a means of prolonging shopping visits. This is not self-evident.

On 6 January 1996 the *Estates Gazette* reported a visit by Hammerson's retail director in the UK to her counterpart in Canada in order to develop new ideas from the example of Hammerson's big mall in Toronto. Plans were also announced to investigate other Hammerson malls in France and Germany. Clearly then transnational ownership is likely to promote transnational trends, and many of the changes in Brent Cross may therefore be seen as part of international shifts in management philosophy. Brent Cross might represent merely international trends. But the second article shows a recognition that locality can still mean difference: 'Shop fronts in this country can be boring. In the States, it can go way over the top. It's trying to get a balance' (*Estates Gazette* 16 January 1996). In understanding this phenomenon we, as also the managers, will have to determine what is a reflection of international trends and what is the outcome of the specific position and consumption of the individual shopping centres.

In this chapter we will try and explain why Miller's dreams of a 'future point of nostalgia' were dashed in favour of an overwhelming imperative called 'natural light'. In order to understand these changes we first have to establish that these changes are anything but natural. This can easily be demonstrated through a brief resume of the history of modern architectural styles. Over several decades Britain has seen the development of a popular movement towards a suburban ideal, battling against an official culture of elites which has attempted to dismiss the suburban as vulgar and impose various versions of modernity in both residential and public architecture. Whether with respect to the concrete and glass of the council estate tower block, new government offices, or the supposed 'genius' of Henry Moore and other designated artists whose work is positioned in public parks, the imposition of official modernism has been an establishment creation often received with sullen dislike by the public at large. Popular taste for decoration and facade was despised as vulgar, and forms of modernism including that which with good reason was termed brutalism held sway for several decades. Eventually, however, the vernacular was resurrected under the official title of 'postmodernism' which often copied the commercial sector which

115

had necessarily retained a more populist feel. In many cases the 'common' has eventually triumphed over that which despised it. Even today there is still an occasional modernist flourish from the arts world such as the Sainsbury's supermarket in Camden Town, North London which asserted itself against the democratising and commercial imperatives which favour compromise over principle.

From its inception, Brent Cross attempted a much-softened version of modernism, and one which could in 1976 still collude with the public's positive appraisal of a compromise between innovation and the resonance of heritage. Brent Cross was the very first shopping mall in Britain and was entitled to evoke a sense of progression towards better things, while keeping an eye on concerns about a loss of the past. Its enclosed space and crude exterior spoke to modernity while its fountain and fake stained-glass dome spoke to traditions of large-scale public space. By the 1990s, by contrast, it was not merely an architectural fashion that was in decline but so too was almost the entire ideological substructure upon which that fashion was built. The very notion of modernity with its sense of onwards and upwards has been thoroughly dismantled while ecology takes over as the popular science orientated towards conservation and a distrust of the artificial.

There is then a fairly simple tale that can be told as the background to current changes, a tale of the rise and fall of modernism. But when we try and apply this story to the actual changes taking place, we find a much more complex and contradictory set of events which cannot at all easily be reconciled with this story. Already the 'global' story is contradicted by 'local' trajectories. The current changes in Brent Cross include a massive expansion of sheet glass to cover the entire roofing area. But this same sheet glass was one of the most emblematic features of architectural modernism covering office blocks, council estates and commercial atriums. Indeed at the very moment that Brent Cross is investing money in glass sheets as a means to appear closer to nature, the major supermarkets in England seem to be moving in precisely the opposite direction. The supermarkets can be seen to be giving up on modernist glass-based architecture and instead returning to gabled tiled roofs, preferably with clock-towers. Whether out of town or suburban it is hard to escape the ubiquity of these fake 'village green' commercial complexes, with their own version of nature, in this case emergent from a context of 'olde worlde' Englishness.

If anything the original decorative order of Brent Cross can now be seen to have been before its time. Long before the concept of postmodernism had become the watchword of contemporary architecture Brent Cross was built, centred on a impressive dome of fake stained glass, that stood as a wonderful example of postmodern appropriation of a generic sense of heritage. So at one level it seems that Brent Cross has just spent a fortune demolishing a prime example of early postmodern architecture and replacing it with something emblematic of now much dated modernism. It is immediately evident that the

changes taking place cannot simply be read off some general or global characterisation of the zeitgeist glibly summarised by terms such as the postmodern,
which are intended to incorporate all current trends. This is partly because a
current trend back to nature or ecology does not designate any objective criteria.
Rather we are dealing with a much more complex phenomenon which is the
public perception of what is natural at a given time. To demonstrate the degree
to which this perception fails to accord with any given property of the architectural forms in question requires that we cross over the North Circular road
from Brent Cross to the shops that are found on the other side.

Within a short distance of Brent Cross lie two enormous 'sheds', vast
enclosed spaces with little or no natural light. One is filled with toys and
called Toys'Я'Us; the other is filled with organic foods and is called Community
Wholefoods. They are of a similar size. In both cases products are piled high
with rather less attention to the niceties and aesthetics of presentation than is
common in shops. Community Wholefoods is in fact not officially a shop and
its customers are supposed to be only businesses, but in practice a large number
of middle-class consumers who prefer wholefood or are vegan have somehow
managed to becomes 'businesses' in order to shop in Community Wholefoods.
The key point of interest here, however, is that during the ethnography
Toys'Я'Us was almost universally experienced as intensely claustrophobic. There
is a whole class of shoppers who really can't wait to leave the site and who
explain their feelings in terms of atmosphere – that is the sense of being in
an enclosed space without natural light and a kind of airlessness that they find
hard to breathe. Curiously the same shed with the same physical attributes
further down the road, selling wholefoods, evokes none of these experiences.
Despite being if anything darker and potentially more claustrophobic the experience of Community Wholefoods does not evoke any of the concerns that are
constantly attributed to its close cousin, the other shed further up the road.

This suggests that the concept of nature may be as fluid and contingent as
that of modernity. There is plenty of disputation amongst historians as to
whether there is a specifically English conception of nature and if so when it
arose (contrast Thomas 1983 with Macfarlane 1987). Britain has experienced
a quite particular development in its relationship between the country and the
city (Williams 1973). In France, for example, a higher proportion of the population either have connections with agriculture or retain a family home within
the countryside which is closely associated with lines of kin descent. For urban
dwellers the key relationship to the transformation of nature is probably through
cuisine and takes place in the kitchen. In Britain, by contrast, the dichotomy
between the rural and urban is in many respects much more severe, and the
response has been to develop a more particular sense of suburban identity,
characterised by the house with its garden (for this argument see Chevalier
1997). What the larger history of the concept of nature suggests is that we
should not be at all surprised to find that it is primarily a form for the reaf-
firmation of social order (Evernden 1992).

This is why we cannot hope to understand the current changes taking place in these shopping centres merely from a knowledge of the history and effects of the architecture, nor from assuming a given context which can be termed nature. Instead this chapter attempts a direct confrontation with the evidence for the perceptions of shoppers in the context of what the *Estates Gazette* has already revealed about the perceptions of the shop managers. In coming to the shoppers we no longer expect some consistent or clear image of either shopping or nature; instead we are faced with a series of overlapping terrains within each of which these terms gain particular meanings and evocations that are brought to bear on the architectural transformation of the centres. It is only through the kind of intensive work which is represented by this study that we may be able to discern how precisely the experience of shopping in shopping centres has plural connotations for the shopper that produce the actual conceptualisations of nature and modernity that they employ.

Once we have a better grasp of the shopper's perception of these terms we may come to one of several conclusions. It may be that these changes were 'inevitable' if the company wanted to retain a commercial sensitivity to these vague moods and feelings of an aggregate shopping mass that nevertheless become the quite concrete statistics of commercial success and failure. Alternatively, as can be demonstrated with much commercial logic (Miller 1997), despite claims to research and a clear drive for profitability, it may turn out that companies spend vast sums merely following international trends without much idea as to the actual commercial impact of their ventures. As in the case of advertising, capitalist firms spend vastly more money 'just in case it has a positive effect' or because their rivals are spending similar sums, rather than because they are confident of the results of their expenditures.

Turning from the architecture and from management to the shoppers themselves we need to determine what constitutes an experience of shopping that is relatively speaking natural or artificial for the contemporary shoppers. Such a discourse is to be found when listening in to shoppers during their actual shopping trips. These attitudes become most evident when shoppers expressed their representation and relationship to the shop assistant, a figure who turned out to be a much more common topic of conversation than the shop architecture.

The Brent Cross shop assistant

In a series of focus groups carried out with shoppers in York, Campbell and Hewer (personal communication) found a clear consensus with regard to how shop assistants should behave, a consensus that was generally also reflected amongst our own focus groups and the people of the Jay Road area. In essence, the shop assistant should not approach the customer without being requested to do so, nor should they in any manner impede the free browsing of goods by the shopper. Instead the shop assistant should wait until they are requested by the shopper to help. If, however, a shopper requires assistance from a shop assistant it is vital that the latter are available and are as helpful as possible in responding to the various questions and requests of the shopper. Shops were constantly being appraised in terms of the degree to which shop assistants conformed to this model. One of the factors that separated out the two key middle-class sites of John Lewis and Marks & Spencer was precisely this experience of shop assistants. John Lewis on the whole was praised as having assistants who most closely conformed to this idea of a discrete presence which is nevertheless informed and helpful when requested. Marks & Spencer was not seen as having intrusive assistants but they were often seen as few and far between and not nearly as knowledgable about the merchandise when they could be tracked down.

The apparent 'naturalness' of this attitude to shop assistants is best critiqued by reference to a contrast with quite different styles in the relationship between shop assistant and shopper. In the first case we can compare this view with the presence of some 'American' style shops and shop assistants. In the second place we can see a rather different contrast with what might be viewed as a remnant working-class vision of solidarity. Sylvia is one of the most faithful users of Brent Cross. She uses it for both major and minor purchases. She has an academic, social science, background and although she has become increasingly involved in more commercial ventures she retains many of the concerns and intellectual interests generated by her studies. She is also aware that, if not what she would call wealthy, she is quite 'comfortable' and has considerable cosmopolitan experience both from holidays abroad and periods spent living or working abroad.

We (Miller and Sylvia) made three trips together to Brent Cross and it was evident that Sylvia made considerable use of shop assistants and clearly appraised them in terms of how well they had performed. On our first visit she talked for some length about how she wanted to write a letter of complaint to the Early Learning Centre with regard to their staff. For example, she observed that the counter was so full of merchandise and pamphlets that the atmosphere that surrounded the act of paying always seemed particularly cramped and stressful. This then developed into a further letter to be written to IKEA complaining that although that shop has got several things 'right', the sales staff were not at all helpful. For example, when she had wanted a particular kind of table and had a baby with her, the assistant just vaguely pointed in a direction and had to be summoned in order to ascertain exactly where Sylvia should go.

On our third visit to Brent Cross, Sylvia had both her children with her. She had come primarily for a garment from Marks & Spencer. In this case although the first assistant apologised for not knowing the stock the second assistant was quite up to 'John Lewis' standards in explaining how the garment requested was part of the Autumn collection and would not be in stock for a further week or so. Sylvia was particularly surprised at this information since she had seen the garment a week earlier in the Marks & Spencer in Wood Green, although not with the combination of size and button she required. This news somewhat offended her class sensibility which suggested that it would not be at all likely that middle-class Brent Cross would be behind working-class Wood Green in carrying particular stock.

After this disappointment, the next problem was how to go to Waitrose in order to test a claim made by a friend with whom Sylvia had just been on holiday in the Dordogne that Waitrose fruit and vegetables had at least some of the taste that on the whole she felt French fresh produce possessed but English fresh produce lacked. In order to go to Waitrose, she would have to pass near the Disney Store which could become a considerable diversion to her two children. In fact she made it to Waitrose, without the kids spotting what they were missing, but on the way back her daughter inveigled her into a visit to the Disney Store. The staff in this shop present a quite considerable contrast with the shops previously mentioned. The shopper is inevitably met at the entrance by an American accent wishing the shopper something like 'a good day' or 'happy shop'. On this occasion we were held within two minutes of entering the store in a long conversation by a shop assistant dwelling on the merits of the Disney film *Pocahontas*, which although not available to the general public, the staff had been privileged to see in private session. Within a few paces of extracting ourselves from this conversation another assistant took advantage of a mere glance at the notice for a forthcoming release of the *Lion King* video, to launch into details about the special offers available if we booked our video in advance. It took only another two minutes for a third assistant to be about to launch into the very same details about the *Lion King* video, though here we were able to indicate that we had just been treated to this information after which we were left in peace.

Sylvia commented in conversation afterwards that this degree of hard sell was quite unusual even in the Disney Store and I had noted that at the time the eight fresh faced assistants in their bright turquoise and white uniforms slightly outnumbered the customers in the store. Having escaped the clutches of the Disney store assistant, however, we walked straight into The Gap. Sylvia had purchased what she regarded as a particularly comfortable pair of jeans from The Gap on a previous occasion and was keen to obtain another. Almost as soon as she started browsing she was confronted by an assistant who came over and asked her what size she wanted. Sylvia took a short time to respond, since The Gap uses American sizes which require translation. When she responded with both the British and what she thought was the The Gap equivalent, the assistant told her in clearly triumphant tones that she had got her translation wrong and what she really required was a different size. He added that he always enjoyed getting one over on the customer, and almost immediately afterwards noted with pride that he was the most big-headed member of staff.

At one level the assistants at The Gap and Disney appeared extremely different. Disney assistants were dressed as a cross between schoolchildren and cheerleaders and showed extreme deference and cuteness in their hard sell. The Gap assistant was black, brash and deliberately transgressive in his hard sell. What they clearly had in common was the complete contrast to the Englishness of the John Lewis or Marks & Spencer approach where this kind of forwardness would be seen as an anathema. Sylvia took this in her stride, indeed soon afterwards she took me on one side and explained to me that these were typically American approaches. She knew that she was much more familiar with the United States than I was and so took the opportunity to demonstrate this knowledge, concluding that the character and style of selling we had encountered was simply a part of the more general sales pitch of both Disney and The Gap, which intended to represent themselves as explicitly American retail outlets thereby standing out within British high streets and shopping centres.

As a result Sylvia had no problem in accepting such aggressive sales techniques as long as they were couched within a framework of national difference that amongst other things reaffirmed that naturalness of what is now the British approach by virtue of an explicit contrast with this foreign style. Not everyone, however, could abstract this knowledge or sense of distinction in order to feel comfortable with both modes. Other shoppers resented the Disney approach and saw it as corrupting of their children (a theme to which we will return to below). The problem of different national styles was felt particularly acutely by Constance, an American living close to Jay Road and married to an Englishman. While he was an academic, she had worked in retail in both England and the States, and could not as yet comprehend the English version of the relationship.

Constance had worked both in a West End department store and in a small shop in Ibis Pond. She suggested that

> in the States people will tend to be a bit friendlier and less guarded. They're more willing to chat and they're used to people helping them in shops. They're used to you saying – Hi can I help you? – they don't mind being approached. They don't think you're necessarily going to sell them something that they don't want and they're more likely to have a chat with you.

Questioned about approaching customers in England, she replied:

> You have to be really careful the way you do it because they just don't like it. A lot of people would just as soon walk out of a shop than have someone come up and talk to them. They seem to think you're going to try and sell them something that they don't want and I feel like, well I'm not going to pull your cheque book out of your bag and force you, and I don't know if they feel they're not strong enough to resist or who knows?

Constance went on from relating this particular difference in the culture of selling to more general differences that made English social relations peculiar to her. She went on to observe that similar differences are evident in her relationship to her neighbours to those within the shops. While she felt that in the United States people feel free to engage in considerable and open dialogue with people they hardly know, the people on this street (off Jay Road) seem highly guarded and prefer to remain within very formal genres of street conversation, such that she feels 'knows' very few people in the area despite several years' residence. This, as she noted, seems to be particularly true of the more working-class sections of her neighbourhood. It is an observation that overlaps with the general development of a sense of 'Englishness' that was noted in the previous chapter.

In her interpretation of this defensiveness Constance draws attention to what I have evoked in my choice of the term 'hard sell'. This term may be defended as characterising the British perspective of an intrusiveness in the American style of retail interaction. In effect they do not see the enquiry made by shop assistants such as Constance as genuine or natural, but rather an artefact of the commercial context in which underlying the exchange is the fact that the assistant's job is to sell goods. As a result they find the shop assistant's instant *bonhomie* to be artificial and false. In many ways this may not be so distant from the sense of discomfort they feel about the over-effusive and intrusive neighbours. As studies of neighbourhoods have shown, there is considerable suspicion about the true intentions of the 'nosy' neighbour.

The relationship between shop assistant and customer is therefore increasingly understood in Britain as having a particular form, in which the enquiring assistant is regarded as false and making the customer uncomfortable, even though this is understood as 'natural' to Americans. Before we come to the conclusion that this is to do with some deep-seated difference in cultural traditions, we may briefly touch on the local history of the shop assistant. Although this is nowhere acknowledged by the shoppers themselves, there is considerable historical evidence that the highly intrusive mode of shop assistant enquiry was merely a few decades before just as characteristic of British as of American shopping. British retail was traditionally based on personal service. In the classic tailors shops such as Burton's there was no merchandise on view: the customer was approached by an assistant who brought in material to show where and when it was required. The customer then stood stock still as they were measured and introduced to the possibilities. Indeed the very idea of self-service for food shopping was originally seen as something rather vulgar and specifically American when it spread into Britain during the 1960s, replacing the personal service of the corner shop. The main alternative form of provisioning, the street market, doesn't exactly bring to mind an historical image of the unobtrusive sales person. About the only longer term precedent was in the more up-market department stores and even here Britain was much slower in developing this norm than its continental rivals (Lancaster 1995: 69; see also Mort 1996: 134–45, for Britain; and Benson 1994 and Reekie 1993 for comparative studies of shop assistants).

So the ethnographic and focus group accounts only make sense in the light of some extraordinary processes of collective amnesia about the relationship between customer and assistant in English retail. If we want to understand the basis for what is seen as 'natural' today the one thing that will not help us is historical knowledge. Rather we need to turn to a distinctly modern and rapidly developing reconceptualisation of what is being projected backwards as having been once the traditional 'community', or what can be termed the imagined community of creative nostalgia.

Wood Green and Ibis Pond

To explore this question of nature and the artificial in terms of visions of community there is a better contrast to be found within the fieldwork than that located at Brent Cross. This is the contrast between Wood Green and another shopping area called here Ibis Pond. A number of middle-class shoppers saw a distinct difference between these two areas which, as the following quotes make clear, is primarily based on a combination of the notion that Ibis Pond is more of a community and that it evokes a sense of nostalgia:

> I'm not very keen on the Wood Green shopping area either. *[Why?]* They're too modern for me, I tend to like more old-fashioned things, individual shops selling their individual produce. My favourite shop in Ibis Pond is the delicatessen, I love it, it's very old-fashioned.

or

> I think it is sort of a community. Well, I think it happens more in Ibis Pond that I meet people I know on the street. I don't know why that doesn't happen in Wood Green because most of the people I know also tend to go to Wood Green but I think it's because of the layout of Ibis Pond. It's just one footpath whereas in Wood Green it's more in and out of large shopping centres. But I seem to meet more people that I know in Ibis Pond when I'm shopping. I suppose it reminds me in a way of Ireland. I grew up in a small town. You're kind of walking down the street shopping and you meet people you know. *[When you meet people do you just say hello or have long conversations?]* Not more than a few minutes. And there's an old shop up there – it's a very old-fashioned shop – I like going in there, I often go in there, especially just to buy, say, raisins – they have really nice raisins so I buy those for Rachel. Or Christmas time I buy, you know, fruit to make Christmas cakes with. It's cheaper in the supermarkets but I feel, you know, it's a family business and it's just a nice atmosphere in the shop and I'd pay more to buy the stuff there rather than go to the supermarket.

The shop referred to in both of these interviews looks as though it comes straight out of a sepia photo of shops in Victorian England and has the smells and associations to match. Equally there is now a thriving coffee shop recently opened which almost at once became highly popular, but it is one that is situated within a context of goods which evoke a sense of 'olde worlde' atmosphere and ambience. Indeed this is the key problem for many of the middle-class shoppers, that one senses that they would like to identify positively with Ibis Pond as redolent of some lost community, or argue that it retains such a sense of community. In practice, however, most of these households are financially simply not in the same league as the core Ibis Pond shopper, and the sheer cost of goods there and pretensions of its inhabitants (some of whom are well-known media figures) prevents them from asserting this degree of affinity.

This tension was directly expressed by some other regular shoppers in Ibis Pond:

> I think there is a community spirit in Ibis Pond actually. Yes, I think there is. But I personally would like to avoid – I avoid Ibis Pond on Saturdays which is when the community spirit is there because I feel I don't know any of those people there. I've been here 24 years, why don't they know me?

or

> Yes. They're, well, I sort of describe them as – oh dear, this is going to sound awful – um, upper-middle class left but with a certain sort of arrogance as well. I mean the Ibis Pondian I'm describing I can't say I really like. They're the sort of people that barge out of shop doorways when you're trying to get in and they don't even sort of acknowledge that you've held the door for them. Or the other day I was trying to get into the doctor's surgery and I'd got Sophie in the buggy and the doctor's surgery's got two sets of doors, almost directly behind each other, and it's really difficult to get the buggy in while holding one door and trying to open the other. And two people came out and they didn't hold the door. They could see me waiting to go in. And then I was standing there and two people came like round past me and walked straight in front of me and just walked away and let the door swing back, you know. And they were, not the two that came out but the two that went in were, you know, sort of archetypal Ibis Pondian . . . Sort of quite arrogant sort of people. But they look terribly smug. They look like they sort of think that, you know, everything that they do is very politically correct and right and everything but they kind of ride roughshod over people too. They're not the sort of people I really like. But then again I do like Ibis Pond in general.

Wood Green is very rarely spoken of in the same tone of community whether idealised or ideals-betrayed. Although some people strongly and positively associate themselves with the area, it is rare that it is abstracted as an image of either what or how a community should be. Instead the identification with Wood Green often takes a rather more pragmatic form. For the working-class population of the area around Jay Road there were relatively few job opportunities locally. By far the dominant niche for earning a wage, especially for women, has been working in retail, most often as a sales assistant. Many of the working-class people in the area had worked in shops at some period in their lives, and this seemed to be particularly true of the more elderly people of the area. Possibly as a result, although other factors may also have played a role, they showed an entirely different point of orientation to Wood Green which was through personal contact with actual sales staff in the area. Many of them had made friends with individual sales staff, or knew someone who was now working in sales. One of the most regular shoppers in the area amongst the elderly was typical in that once a week she would meet a member of the sales staff at British Home Stores and have lunch with her.

Perhaps the most extreme example of this form of affinity came from an informant who lived at some distance from Jay Road but became an informant since she had her hair done at a salon there. An elderly Jewish lady, she clearly identified with Wood Green as her locality, and as with many others had spent most of her life working in retail. It was clear that in retirement she found it difficult to give up an activity which had been so much part of her identity. As such she found the market square within Shopping City to be the most convivial area to visit, because in some ways it was more under the control of sales staff than would be the case in more formal shops.

From our discussion it became clear that her affinity with the area went well beyond merely having a chat with sales staff:

> I've been there so many times I know quite a lot of people that work there. *[Do you?]* Oh yes, quite a lot yes. I know somebody that's got, there's a handbag shop there, and he's got the jewellers on the corner, I won't tell you how many years I've known him! Years I've known him and he works in there and there's some young woman that I know who sells millinery she works in there. I know quite a lot of people ... Well if they're busy I can't sit and chat to them but sometimes if they say can you stand in for a minute I want to go up and get a cup of coffee and come down I'll stand there for a minute. If anybody comes I'll serve them, which I've done I've done that more than once. *[Which shops?]* Quite a few, yes I don't mind it, I've sold purses and suitcases it doesn't make any difference to me.

This could go even further, as emerged in an anecdote about a friend of hers:

> Well she wasn't actually working. She's 83 years old so she only used to go in there. She never got paid for it. She used to go and help this person

125

out for years, quite a number of years until unfortunately he sold out and that was it . . . Well she used to go in there quite a lot and then she used to go in there to help because there used to be a young lady there on her own, and she used to say 'would you mind helping with this I want to go upstairs and buy a few things' or 'I want to go and make a cup of tea' or something like that and that's how it started, and when the governor came in he didn't know who it was, he had no idea, and he said 'What are you doing?' but she was a very, very, very good sales lady because when this person used to go to have a cup of tea or lunch she sold more in that half hour or hour than this person sold all day.

This is perhaps an extreme example, but it stands for a much larger generalisation which holds for many informants. Many middle-class informants identified with Ibis Pond as an imagined ideal community redolent with nostalgia which assumed how communities used to be, though bolstered also by the genuine experience of meeting other shoppers that they knew. In the case of Wood Green this was replaced with a community of personal association built up largely through the identification of working-class people with those who continue to work as shop assistants. The symmetry of the two class positions becomes more acute when the relationship between shoppers and shop assistants in Ibis Pond is taken into account.

A question asked of most informants was whether they had come to know any of the shop assistants in the areas where they shopped. It soon became clear that this was highly unlikely for middle-class shoppers shopping in general. But the degree to which middle-class shoppers distance themselves from the shop assistants was made much clearer by two of the informants who had themselves worked at some time or other in Ibis Pond where most shoppers are middle class. One had worked there in a bakery:

> I used to work in Ibis Pond, I don't like Ibis Pond, I find it very expensive and there's not really much up there for me. *[What else don't you like?]* People, I just don't like the people I think they're too above themselves . . . They're very above themselves . . . the bakery I was working in was very busy. They come in and you'd give them the change, it wasn't good enough to hand them the change in their hand they expected it to be counted penny to penny and like your attention constantly had to be on them regardless that the shop was absolutely full of people, you know always have, it's very difficult to explain but like in Wood Green it was fine you know people aren't like that they know they understand. *[Wood Green customers?]* Well they're good people, they're working-class people. Well what can you say? You're all in the same boat, so to say we're more . . . Like it wasn't like a customer, basically you had a lot of friends from all the shops, people became very friendly. *[Are you still friendly with them?]* Yes, I still go in Wood Green, in many shops. They're still working there.

Always 'Sheila, come on, when was the last time we saw you, you don't come in and say hello any more'. All this stuff.

The same point was made by the other informant who had worked in a shop in Ibis Pond:

You get a lot of women who sort of expect the earth for nothing sort of thing, and it's your general 'let's look down on shopkeepers' attitude. You get a bit of that and it might just be that Ibis Pond it's a bit pricey to live there.

Natural values

Ibis Pond clearly evokes a series of tensions within (mainly middle-class) people's ideas of community. This should be something which exists between real people expressed in their degree of sociality. As such it should also connote a past time when such sociality is deemed to have been the true condition of rural life, which is now in danger of being lost. The tension arises because the character of Ibis Pond as an up-market shopping area at the same time evokes precisely that factor which is usually blamed for the loss of traditional community values. In particular there is a direct relationship drawn between the wealth of the people who live there, their materialism and their anti-social nature. It is the association with money itself that is constantly juxtaposed in people's conception of what has led to their current indifference to ordinary human civility and concern with others. The sentiments that are to be found in the everyday discourse of most shoppers echo many of the more academic comments on the destructive effects upon communities of money as an abstract form, for example Simmel writing about Western Europe but also in commenting upon many ethnographic contexts (see Parry and Bloch 1989; Carrier 1995). By contrast, Wood Green is viewed as naturally civil precisely because the comparative poverty of the people in the area grants them a kind of authenticity with respect to ideals of community.

Such constellations of values incorporating ideas of time, wealth, morality and society tend to involve many basic distinctions that resonate one with another. For example, in conversation there is a constant affirmation of children and childhood as also expressing a kind of original authenticity which may be corrupted by materialism and the artificial (see Miller 1994). This in turn evokes the nostalgia that people express about certain shops in Ibis Pond as being like the shops one used to know (or more likely know about) and may therefore be associated with romantic images of one's own childhood. About the most violent scene witnessed when actually out shopping was when the same individual who is talking above about the arrogance of people in Ibis Pond came into the special unit which had been set up in Brent Cross to answer people's queries about the current new developments occurring at the shopping centre. She started by suggesting that the new developments should include a crèche area where children could play but with an additional barrier so toddlers could not exit by themselves unnoticed by mothers. She then became extremely agitated when she found out, in the first place, that although such a thing was being considered, no decision had been made, and then in answer to a further question was informed that to date most of the customers' suggestions had been with regard to car parking and the bus depot rather than children. She harangued the member of staff concerned quite vehemently about how wrong this was. She was not soothed by the staff member noting that as staff she only recorded the complaints and could hardly be blamed personally for the views being expressed by the general public.

Given that there is relatively little danger to infants in Brent Cross as against in the High Street, it is possible that much of this concern (which was also echoed by other shoppers) is more to protect their children from materialism (as in the ambivalence about the Disney store already noted) and, in particular, what comes across as the children's own unmediated materialism. That is to say it is not only the sense of corruption of the child by commerce that is at stake, but a kind of embarrassment that their child's innocence is manifested in the form of unmitigated desire and greed. Given that so many of our findings about adult shopping reveal a complex striving to carefully modulate and modify their own desires for goods within what they could come to regard as acceptable moral frameworks, it may well be that the child in a sense exposes them to the fragility of their own strategies for taming desire. It is within this context that we should have little difficulty understanding why the same individual who feels claustrophobic and unable to breathe within the shed occupied by Toys'Я'Us, where the image is constantly of children swamped by goods, manages to have quite benign feelings about the equally cavernous and enclosed shed further up the road that sells organic foods.

An equally common idiom within which these distinctions could be expressed arose from the idea that some goods are themselves more natural than others. In general shoppers who wanted to oppose the evils of materialism as antisocial could do so through a preference for goods that were somehow organic, wholesome and themselves closer to nature. This in turn could be turned back into the dimension of time as organic, associated with traditional and past values. These values can also be expressed without abstaining from what are in practice more traditional foods such as meat. As one from the same group of shoppers, noted, 'I suppose we don't eat much red meat at all. Really I would look, I buy sausages – that's the nearest to red meat we get – I look for the ones that are the traditional, sort of hand-made, sort of thing.' Although in practice the actual work done by shoppers in support of green or ecological concerns was relatively slight, the interviews suggested that this remained an important part of their self-conception of their responsibilities as shoppers.

Many other values and concerns could be brought to bear upon this discussion but that is itself the main point that is made by the previous argument.

The concept of nature as used in relation to shopping is not one which can be isolated as an attribute of architecture, or as an evocation of some ideal state prior to the rise of human artifice. Rather it is a term which brings together a wide constellation of values which can then be applied to an almost infinite range of new circumstance. It is best treated as an aspect of what the anthropologist Bourdieu (1968) called 'habitus', a set of foundational ideological premises that lead towards a general disposition which is manifested as the holder of these values encounters both familiar and novel settings about which they need to form an opinion. It is at this level that attitudes to children, to Englishness, to food, to holidays and leisure activities, to community, to taste, to ideals of sociality all come together.

A final example may help to make the point, since it is one in which feelings about children and materialism become involved in what has been the main exemplification of nature and artifice used here – that is, attitudes to shop assistants. One of the wealthier shoppers was reflecting on an experience within a children's clothes shop in Ibis Pond:

> It was about £80 or something like that. So therefore these days I don't go there. But the women there are so nice, I mean they are sort of so friendly and they know how to – I mean they sell a lot because they have a way of actually relating to you, making you feel one of them, sort of thing, you know, and in the end you buy one. Because I can't, but somehow they make you feel that you could, you could if you wanted, why not, sort of thing.

Here the shop assistant's very sociability acts to allow a shift towards greater indulgence because the degree to which that indulgence might be experienced as anti-social is assuaged by this context of an apparent social relationship. It is with this sense of the complexity and multi-faceted nature of shoppers' values that we may return to the original question of architectural changes in shopping centres, no longer expecting to find either consistency or changes that can be read off from the basic trends in architecture over the last few decades. Instead it is a world of constant ambivalence and arguments, where one person's compromise is another person's hypocrisy. Or, more concretely, where for one person the Body Shop is an exploitation of nature in the interests of commerce, for the next shopper it demonstrates that one can have commercial success while marking genuine progress in our sensitivity to animal and human welfare.

It follows also that the trend towards 'nature' cannot be separated into actual differences between more or less natural forms, nor between attitudes to objects as against people. The construction of the shop assistant who does not approach until asked to become a resource as a 'natural' relationship should strike one as no more peculiar than the idea that having glass roofs instead of plastic ones brings nature into the shopping centre. As was evident from the conflict between the British and American view it is equally reasonable to regard the new British notion of the natural shop assistant as an unacceptable dehumanising of workers as it is to see it as a growing acknowledgment of the 'truth' of a relationship in which one is being approached principally so that one can be sold goods.

A great deal of the recent literature on shopping centres is focused upon the nature of fantasy, influenced both by the writing of Walter Benjamin and more recent general debates about postmodernism. The characteristic contemporary article on shopping malls tends to focus on the development of shop fronts as tableaux often closely connected with the façades found in the major tourist destinations or with key historical epochs (Chaney 1990; Hopkins 1991; Shields 1992; Williamson 1992). These new façades are very evident in the most recent British shopping malls, such as the Gateshead Metro Centre and Sheffield's Meadowhall, though it is not found in either of the centres we studied. Most commentators see such façades as evidence for a new superficiality in the relationship between shoppers and the task of shopping and see them as emblematic of the set of relationships that have been characterised as postmodern.

We have not undertaken direct studies of the way shoppers perceive and use such façades since they are not present in Brent Cross and Wood Green, but we may be able to extrapolate to some degree by looking at the relationship of shoppers to the changes in façade that are the topic of the present chapter (compare also Miller 1997: 268–74). The premise of our argument is very different. It does not deny that such changes may to a degree evoke the relative freedom of expenditure which most people associate with the sense of being a tourist on holiday, or that this may be of benefit to commerce which would then account for the trend. But while this explains the commercial motivation it does not fully address the reception of such innovations by the public, where we would suggest it becomes part of the larger relationship between ideals and reality that are played out during shopping. For the shopper these changes in façade – so far from being superficial – relate to the very profound ambivalence that shoppers feel about a whole constellation of values. Because these values bear on immediate questions such as the proper form of parenting, the amount they should spend, the ethics of conservation, etc., they cannot be held to constitute 'fantasy' where that term is opposed to 'reality'. Such a view would have to assume that such issues of representation and symbolic forms taken in relation to ambivalence about social issues was a peculiarly modern problem while older societies were somehow more functional and utilitarian. Yet the crux of Lévi-Strauss's (1972) analysis of myth was precisely that it is a collective representation dealing with ambivalence caused by contradictions in social and moral relations.

Accounting for the changes

With these strictures in mind we can have another go at accounting for the articulation between management and shoppers' perceptions of the changes that have been made to the centres. We have argued that, on the whole, working-class shoppers in this area have a more direct relationship to the shop as a location of work. This is not because they are in all respects more grounded and sensual in the way Bourdieu (1984) tried to imply for the working class, but for the simple reason that they have themselves considerable experience in retail which represents the dominant form of local employment in this part of North London, as it has done for a considerable time. For as du Gay (1996) has argued, this is an area where production and consumption identities have fluid boundaries. Since either they or close friends or relatives may have direct experience of retail as a site of work, it is not surprising that their immediate point of identification is often through the people who serve in shops. This in turn constitutes the main link they construct to the shop as a site of sociality.

The middle-class shopper, by and large, apart from some possible holiday job while at school or university, has very little experience and virtually no sense of identification with the shop as a place of work. They are rather more concerned to use the ambience of the shop to resolve the constant sense of ambivalence and contradiction which they experience because of an almost obsessive concern with various idealised moral universes. In particular, they are concerned with the conflict between an idealised notion of community and the idea that materialism has been the main force which destroyed this mythical state. In this particular case this opposition is diffused through a number of other moral dimensions, of which the most pertinent is a longstanding contradiction between an appreciation of the advantages of artificial construction symbolised by modernism and a fear for the loss of an equally idealised state of nature which again is taken to be destroyed by the relentless rise of commoditisation.

This argument, and indeed this chapter, rests on the assumption that the changes that were being made to Brent Cross were addressing an ideological contradiction rather than a problem that is conceived of as essentially functional. Such a conclusion seems consistent with the generality of comments made during the research, whether these arose from the focus groups or the ethnography. There are not that many comments from either source that directly concern any potential functional implications of opening up to natural light, partly, of course, because the functional changes are actually quite limited. Where comments are made they are as likely to oppose as to welcome such changes. More comments pertain to the nature of the enclosed space than to the view of natural light. In fact the arrival of natural light does not of itself lessen the degree of enclosure. Comments are made such as 'there's no actual fresh air, it's just so crowded all the time', 'nobody realises just how far you're walking in the shopping centre', 'it's hot, it's uncomfortable, it seems like everything's crammed together'. There are, however, at least as many comments that welcome the artificial environment thereby created: 'in the summer when it's really, really hot, in Brent Cross it's cool and in the winter when it's really, really cold, it's also a nice temperature', 'that's why I like Wood Green 'cos

131

it's enclosed and once you're in there, that's it, you don't have to worry about the weather, you don't care whether it's sunny or rainy outside or whatever'.

Even if one takes the specific act of opening up to natural light there are disagreements. On the one hand, 'I think it will be less claustrophobic if you let natural light in', 'I don't mind when I'm in there, but I have to get out after an hour or two, that's enough, and then you need to see daylight'. On the other hand, a shopper commented on the potential changes:

> Oh God, as long as it's not natural sunlight at the moment, because one of the things we did find out last year when we went to Brent Cross on an incredibly hot day and thought this is really silly going shopping on a hot day, and it was wonderful, because everywhere was air-conditioned and it was so comfortable. We came out at sort of quarter to six and of course it was boiling hot outside and we thought oh we've been in the most comfortable place. It was very nice actually.

Indeed even the officials concerned with the change do not seem to see any functional necessity behind these moves. Rather, they emphasise the idea of keeping up with the times. They have the sense that Brent Cross and Wood Green were looking 'dated' since all the new international shopping malls were being constructed with natural light, and this seems to have been the major concern. Wood Green has moved in this direction but to a far lesser extent than Brent Cross. The managers clearly saw that the concern for nature as something resonant of the lost past was simultaneously a desire to be and to remain modern. It therefore seems reasonable to take these changes as part of a larger movement which makes opening up to natural light more a symbol of letting nature in to ameliorate the concern of shoppers that the act of shopping might itself represent an un- or anti-natural activity. The shoppers do not have a consistent view as to the functional benefits or deficits of the change, but the changes do fit within a more general relationship between ambience and their changing perception of nature. On the whole, as in so much of middle-class life, the concern seems to be not to express any clear principle but to devise strategies that engineer compromises that allow people to live with ambivalence, based on their understanding of the contradictions that are evident in their lives; the same suburban, semi-detached compromise that is viewed by others as hypocrisy.

One can see why Brent Cross would have more of a problem over the question of artificial and natural light than would Wood Green. Although such concerns are to some extent general to almost the entire population, they are felt particularly acutely by the middle class with their greater exposure to and experience of materialism. Newspapers which are directed to this fraction of the middle class, such as the *Guardian* and the *Observer*, tend to dwell upon these questions of morality and ambivalence when discussing retail development: 'perhaps superstore shopping, for liberals at least, will be like voting

Conservative: a lot of people will do it but none will want to own up to it' (*Observer Life* 27 February 1994; see also the *Guardian* 9 February 1995 and 14 March 1995). To the extent that Brent Cross is the more middle-class site then this is where the issue will arise most clearly. The location of Brent Cross exacerbates the problem, not just being sited off the junction of the North Circular Road and the A41, but mostly because it is cut off as a distinct shopping mall. While out-of-town stores have at least some pretention to balance with their rural settings, it is hard to take the naming of the hotel 'Garden Court' opposite Brent Cross as anything other than a bad joke.

One of the many ways in which shoppers act to ameliorate the negative feelings that they have about shopping as an act of materialism is by situating the shopping act within some other concern. As long as there is some sense of high street or open aspect, then there is a potential ambiguity between going shopping and going for a walk. Amongst the elderly, in particular, the two activities were often elided, so that Wood Green high street became essentially the site for a 'constitutional'. This may also be one of the reasons why shoppers tend to ignore many of the differences between Shopping City and the high road, assimilating the former within the latter. In this respect Shopping City may also be helped by the presence of the market-place which for some shoppers at least helps to evoke a sense of a different, more open shopping environment. Indeed, for this reason Wood Green can probably afford to draw direct attention to its enclosed space by advertisements which focus upon the attraction of getting out of the rain by popping into the centre.

None of this is possible with Brent Cross, where shopping is unequivocally a shopping expedition. It would require quite some ingenuity to convince oneself that one is there for any purpose other than shopping, an act which is expressive of the materialism of the shopper. For this reason the shopper at Brent Cross has much more difficulty affecting the kind of compromise and balance which are central to middle-class strategies of value creation. It is they who have the discretionary income and can influence the aesthetics by the power of demand. Although there are many positive features that can be associated with compromise the tendency to ignore contradiction also has its malign side, as has been made evident with respect to the relations to workers. The middle class are found to be so busy constructing ideal models of community that they become blind to their own disdain for the bulwark of any actual community amongst the people who labour on their behalf.

We are not arguing that shoppers are not quite open and content with the idea of themselves as shoppers and the potential pleasure in the activity of shopping. There was plenty of evidence from this research that shows this positive relationship to the act of shopping. The point is rather that they simultaneously want to think of themselves as sensitive to the negative implications and connotations of that aspect of their identity, and also be associated with a concern with nature and the potential evils of materialism for society in general and themselves and their families in particular. Sometimes this contradiction is explicit, as with the development worker who when asked about her shopping stated, 'Every two weeks. I love Brent Cross. I mean I hate the fact that I love it and I hate being such a – but I do go.' More commonly these contradictions are embedded in a wide range of discourses which, as argued in this chapter, range from authentic society through natural children to foreign-style shop assistants. It is precisely because these contradictions are diffused within a general constellation of values and contexts that activities such as opening up to natural light which affect such a vague and diffuse attribute as 'ambience' can be extremely important to the success of a shopping centre.

To conclude, there are several reasons why supermarkets and Brent Cross can be going in opposite directions in their pursuit of nature. It is hard to find persuasive functional explanations. At best one might argue that supermarkets are viewed more from the exterior, while at Brent Cross the experience is more from the interior, but this would be hard to demonstrate and is probably only marginally, if at all, the case. More important is the actual historical trajectory of each site. Put simply: if Brent Cross had been built of modernist glass in 1976 it would very likely be moving towards the postmodern today. Since it was already postmodern it has had to use a modernist motif. This makes sense in the light of all that has been shown in the chapter about the extremely dynamic character of the perception of 'nature', such as the total reversal of ideas about customer to shop assistant relationship within two decades. As many anthropologists have shown, ideas of tradition and nostalgia emerge as the other side of the coin to ideas of modernity and are just as dynamic (Rowlands 1995). The shopper may be able to think of nature as a stable category that they are always in danger of losing. But the managers have to recognise that this category is actually extremely dynamic. So that to remain in touch with nature will demand a constant shift in retail image reflecting these changes in the consumers' concept of nature. As in so many other areas of contemporary life, the problem with perceptions of nature and tradition is not that they are so old and stable but that they are so new and dynamic.

7

JOHN LEWIS AND THE CHEAPJACK: A STUDY OF CLASS AND IDENTITY

Introduction

To use the term 'class' in a chapter heading could potentially obfuscate rather than clarify the intentions of that chapter, since the term has developed a wide range of both colloquial and academic denotations. We wish to start, therefore, by specifying which particular sense of the term is being employed here and with reference to which section of the relevant literature. Crompton (1993) provides a useful summary of the current array of positions with regard to sociological approaches to the topic. Within this she defends the continued validity of class as an analytical and descriptive construct.

The ethnography was deliberately sited in a road where it was hoped extremes of differentiation would be minimised. It was felt that many studies take categories such as 'working class' or 'elites' as their intended subject and tend thereby to ground their findings within the same 'given' category that served to site the research. The ethnographic section of this research was designed in order to be able to take a fairly nondescript site which did not lend itself to a priori classification and thereby allow the more complex realities and salient self-designations of contemporary urban identity to emerge from the process of research itself.

In practice, however, it was found that there existed considerable differences in what an outsider might see as the objective life-chances of households. Although no systematic correlations were attempted, observation suggested that these differences were still closely related to both the history of education and occupations of householders. The most marked differences were between those living in council estates on one side of the street and those in private accommodation on the other. In effect, then, Crompton's insistence upon the empirical and analytical vitality of class would most likely have been affirmed if we had carried out a more systematic study of this issue.

It was not, however, the intention of this study to contribute to the tradition of class analysis that Crompton surveys. Most of this literature derives from classic approaches to class and social stratification developed by Marx and Weber. These were initiated within larger theories concerned with the dynamics of power, contradiction and dominance. As it developed within sociological practice, however, the concept of class became increasingly orientated to the more pragmatic problem of examining social classifications in order to determine which social categories (e.g. occupation or housing) might prove useful predictors of behaviour and expectations in other sectors, such as voting or social mobility. The intention of our study was rather different. Our concern starts from a particular relationship between subjects and shopping centres in order to ascertain the possible implication of this relationship for the identity of shoppers.

Crompton notes that more recently there has emerged another approach to class, which at first might seem more relevant to our concerns. This is based on the assumption that class association and affiliation may be decreasingly based upon occupation and increasingly linked to the arena of consumption (see also Burrows and Marsh 1992; Warde 1994). The primary reference point for such studies is the book *Distinction* by Bourdieu (1984). The argument in favour of this approach and subsequent studies such as Savage et al. (1992) on the British middle class is that they take into account a whole swathe of more recent cleavages within social categorisation that appear increasingly pertinent to class position, for example, the differences between those who are 'high' in terms of cultural capital and those who score highly in terms of financial capital. Following Bourdieu, but using a rather more refined methodology, Savage et al. (1992: 99–131) make considerable use of consumption profiles in order to tease out the precise relationships between these different forms of capital within different fragments of the middle class.

Although these studies are clearly much more closely involved in consumption they remain a revision of the larger attempt to construct a series of hierarchical categories as foundational to social analysis. Furthermore, they tend to relate all findings back down to classifications based on occupation and abilities to appropriate or exploit forms of capital. Once again, however, this is not the intention of the present study. Rather than starting from Bourdieu's *Distinction*, it is a separate paper by Bourdieu on the genesis of class (Bourdieu 1985) that would seem a more appropriate starting place from which to set out our findings. Here Bourdieu points out that this extensive discussion of class within sociology and most particularly within Marxist studies has, in effect, radically altered the very phenomena it was intended to analyse. Perhaps more than any other term in social science, class creates the very discourse that it purports to describe. Whatever its basis in social and economic conditions, such as relations of production, we are by now so saturated in its discussion that class has itself become the taxonomic grounds upon which society in turn becomes organised and conscious of itself. People of all classes commonly discourse about themselves in relation to an explicit concept of class. This generates what Bourdieu calls the social space within which the self-categorisation of social groups has to be situated. Given that Marxists in particular were concerned to 'raise consciousness' this is not an unreasonable outcome of at least Marx's theory of class.

136

Such debates provide a starting point for examining the self-reference to class found within an ethnographic enquiry as an act of consciousness. But even this article does not yet reach the point of departure for the current chapter. Bourdieu's concept of class as discourse is one in which people's consciousness develops on the basis of what is now a long history of debate about class that as it were establishes a foundation for self-designation. But once this has come to be, we would argue, populations start to create their own sense of class as the holder of values and operationalise class within their own social strategies, which objectify this general discourse of class through specific practices within their localised environment. As in the Weberian tradition this leads towards a further concern with boundary construction and with legitimation, often as Lamont (1992) has recently shown in her comparative study of upper-middle classes, with powerful moral overtones.

By this stage we become concerned with class as a process of objectification. That is, we can examine a particular medium around which a sense of class is constructed. The premise is that this medium may act both to reveal to social agents something of what class can be – the values it encapsulates – and further to inform them as to who they are in relation to those set of possibilities that have been revealed. Class is then more an act of becoming and understanding of oneself in relation to objectified values than merely an act of situation within a given classification.

This idea of class as process is particularly suited to the underlying topic of relating a particular medium – the two shopping centres – to the topic of identity as discussed elsewhere in this volume. The chapter does not pretend that class is solely an outcome of shopping. Clearly both as analytical category and subjective label it rests on historical discourses and objective conditions of occupation, education and housing, with which the more traditional literature on class has been principally concerned. The present chapter builds on such theory and analysis by contributing a discussion which is tangential though relevant to them. It starts with a coming to consciousness of that which is already given, but moves on to the use of the material world to articulate an element of identity as process.

In order to accomplish these tasks the chapter follows two particular channels of enquiry. In the first place it will attempt to create a dimension along which can be strung informants' views about shopping, expressed both in conversation and in their behaviour while out shopping. It will be shown that this provides evidence for a clear status hierarchy within which people relate themselves to shopping sites. Here then one sees the first strategy – where shopping is used to create relative positions within the group. This may well owe more to their prior sense of class imposed as habitus upon the shops, though the shops then become an additional medium through which this classification is clarified and given further meaning. This will be followed by an example of material culture analysis in which the emphasis is on an analysis of the shops themselves and their ambience in the light of the former dimension.

The analysis of the shops reveals a more structured polarity. One of the main concluding points is that despite the diversity of shops within each site and the similarity between shops in the two centres, the members of the shopping population create a clear sense of polarised opposition which helps them clarify the concept of class to which they in turn relate themselves. The two sections thereby relate to two aspects of class theory. The first treats shopping centres as more reflective of given class identity; in the second half, the active component of class as a process of becoming is the focus.

The background

As already noted the field-site for the ethnography was chosen in order to include a wide range of income groups and prevent the simple designation of the site as a homogeneous social category. We therefore picked a fairly nondescript street (Jay Road). The two council estates (Sparrow Court and the Lark Estate) on one side of the street appeared relatively affluent estates, and likely to contain some households with higher disposable incomes than much of the private housing on the other side of the street, which was dominated by small purpose-built maisonettes. In effect we wanted to centre on something like average national house prices of £60–70,000. As the research developed it also included households in the side roads which are somewhat more affluent. Since informants often do not differentiate between the shopping centre and the high street in Wood Green it is the shopping area as a whole that is at issue here.

If we wished to diminish the sense of class, however, then this selection failed. There are too many people who will account for their actions in terms that, if they do not use the word class (and many of them do), certainly seem to imply that there is a group of people 'like us' who differ from 'them' and are indeed defined by this opposition. Colloquialisms such as wanting to move into an area that would have 'kindred spirits' abound. Although around half the population are either not British born or belong to an identifiable British ethnic minority, we have still found that, be they West Indian, Cypriot or Asian, most households are commonly using a class-based language of similarity and difference. Any doubts about the importance of preserving social differentials would be quickly dispelled by the sheer number of conversations that turn to the issue of secondary school selection and the high level of anxiety this fosters. When it comes to parents discussing their children's schooling, the British class system appears in its most triumphant and insurmountable form.

To note that class is of considerable concern is not to suggest that people are clear as to how they can relate to class. This may have been particularly problematic in the street chosen for fieldwork. As many people on the street commented, this area, and, in particular, the private housing sector, is a place of considerable transience, which meant that it was just as hard for householders as it would be for a sociologist to characterise them in terms of stable class categories. A large section had the kind of educational background that should have led them to reside within the middle-class groups defined by Savage et al. (1992). But they were often quite uncertain as to how far they would be able to translate these into any kind of future capital. This was not merely a reflection of an increasingly insecure job market, but also because there was a relatively high number of women who were looking for work in sectors such as television, acting or journalism where they had relatively little idea of whether they would end up very well off or quite the opposite. Furthermore, the council estate revealed a number of families which suggested downward mobility as children of families that had previously been in owner-occupied dwellings. Finally, this is an area with a large proportion of people born overseas who were still trying to establish where within a class system they would see themselves and were being seen by others. The result, then, is particularly complex and it may not be surprising that the sense of class as a constellation of values is less premised on any given social foundation and rather more on establishing comparative relationships between groups and processes of objectification.

Wood Green and Brent Cross: some positions

In many respects the polarity between Wood Green and Brent Cross is a better indication of any one individual's perspectives than reference to either one of these sites alone.[1] There is a strong correlation between the views on the two centres, which certainly justifies the decision to pick two rather than a single centre. This result is all the more striking given the conclusions in chapter 3 that demonstrate how the populations which use the site are actually rather more similar than we had initially anticipated. We have organised the views expressed into a series of categories. When placed in juxtaposition they express a dimension of difference which equates with a gradation along a sustained sense of status difference. However, as will be made clear in most cases, the category is forged out of a more complex sense of identity in which it incorporates alternative cross-cutting dimensions of distinction.

The sense of being above

We have only one example of the sort of person who clearly feels that Brent Cross is beneath her even though she regularly shops there. She is West Indian and although not particularly well off at present, has been in the past. She lives in her own house in one of the streets near Jay Road. She describes her husband as 'very Mayfair, Sloane Ranger type', 'He doesn't like Brent Cross. He doesn't like north London full stop. He's more such a central person and such a . . . he's such a . . . he's a person who doesn't – he thinks of coming down and he just can't do it'. *[To do with his background?]* 'Actually his background isn't that but I think he's for a certain status and level and he doesn't want to come down.'

She in turn reflects this attitude as is evident when she is discussing shopping for a lamp in the sales:

> [*Did you get anything in Heals?*] Yes I did. They were very expensive. They didn't have very much of a sale in their department. Then I came back to Brent Cross and got that. It wasn't in the sale but obviously if it was I would have got it – the blue one. I got it in John Lewis. I was looking for a specific thing and actually Heals suits my taste better than John Lewis does for these lamp things and I wished that . . . For some reason, it must be the post Christmas thing, but even Heals don't have that much stock that would excite me to spend on it.

Given that where she lives is not a wealthy area it is unlikely that one would find many people who consider Brent Cross so beneath them that they would not shop there at all. She is well aware of this as she notes: 'I'm afraid even the people shopping at Brent Cross are a bit scruffy. You can tell the difference. But of course the people in Brent Cross are very much a part of this area really.'

Even where there is the same shop present she would prefer the West End: 'I think I will have a look, see what Benetton has got in Brent Cross, but I think there's a Benetton in Selfridges. I think I will go to Selfridges and look there because they've got all my – the shops I would buy for her, her clothes, Patricia Wigan's, Bonpoint.' She does see hope for Brent Cross if it's changed in her direction:

> And I wish that they would just upgrade it. But I have seen other people's malls have overtaken our Brent Cross. So I think being the innovator of malls, I think they should now put another floor on it. [*You know they are about to change it?*] What are they going to change about it? [*They're changing it to natural light. They're taking the dome out of the centre.*] Oh, are they? How wonderful. But they should add another floor before they do it. [*What would you like them to add in?*] What, shop-wise? Maybe a Liberty's type thing and – which is general – because we need more than just John Lewis's and Fenwicks and there's a Marks & Spencer's. We need something a bit more pricey and a bit more graded upwards.

Many of the shoppers living in private housing had difficulties relating in any positive manner to Wood Green. Negative feelings about Wood Green, however, almost all took a particular turn. The issue of safety and violence is used to more or less cover any other grounds people might have for not wishing to shop there (see chapter 5). Mostly these are just general remarks, for example an informant who noted: 'I don't relax so much at Wood Green as I know people who have been attacked, one was pushed and had her bag stolen, the other had her purse taken without even knowing it.' Occasionally people will go into more detail about their own experiences. For example:

I hate Wood Green, it has got a lot of racial problems. I waited for my husband, he took the children home in the car and I shopped alone and took the bus back. I was sitting on the bus and there was a gang of stupid teenagers, I was on the bus and he was outside and he spat on the glass, what a nice thing, eh? Everytime I go something like this happens. Once I was in a queue waiting for a photo for £1, there were two teenagers with children in front, one of them had taken the child to the toilet to dress her up in some fancy outfit, when she came back I said 'excuse me but I am queuing' and she turned and hurled abuse at me, I thought she was going to attack me but I didn't know she had just gone – really stroppy people. *[Gives third example.]* It is this violence and hostility which I hate in Wood Green. I am sure it is due to racial or poverty I don't know.

This feeling about Wood Green is by no means confined to those who have an actual negative experience to report. Indeed one informant who identifies closely with Brent Cross noted: 'I feel the whole time there I have to watch my bag and I have to watch when I get my purse out. Its not from any bad experience, but I just, its got that reputation.' The irony in this case was that she does not have the same worries about Brent Cross despite the fact that she then notes: 'My aunt had her purse stolen there last week and I heard someone else in Tesco's say – Oh I had my purse stolen at Brent Cross last week.'

Generalisation can only be tentative but what emerges is a problem of deciding how best to legitimate statements of preference and disdain. There are a number of older residents on the Lark estate who are not reticent about expressing racist views, yet because they identify positively with Wood Green do not discuss this as an issue. By contrast, those who express dislike at the idea of shopping at Wood Green tend to be mothers from the owner-occupied sector who focus clearly on the issue of security but then may express this either in racial or class terms depending upon which of these they feel less embarrassed to blame. The shopper who sees herself as above Brent Cross clearly had little difficulty in expressing her class identity with terms such as Mayfair. Being of West Indian origin she is more equivocal about her dislike of Wood Green. She notes: 'I've been there about five times and I'm not keen on it. But I think I'm now beginning to feel the vibrancy of it . . . you can appreciate the vibrancy of the people in Wood Green where we haven't very much vibrancy and sort of rhythm here.'

It would seem then that to attempt to clearly disaggregate these two issues of class and ethnicity would be to miss the degree of equivocation within the rejection and indeed often intense dislike of Wood Green by some shoppers. But while expressions of dislike are interlinked with other concerns, as it were horizontally, they still suggest that as a series of idioms they come to express a clear vertical position where a particular shopping site lies somewhere 'beneath' the identity of the person concerned.

The pragmatic shopper

A category such as 'pragmatic' does not at first appear to be appropriate within a dimension of status differentiation, yet we believe it is just as much a part of this dimension as the sense of feeling above and below. To take a parallel example, in a previous study of kitchens in North London, it was found that in the advertising literature for fitted kitchens there were three main categories: a nostalgic look associated with items such as pewter and dried flowers; a modernist look associated with spotlights and hi-tech; but between these two came kitchens made of mixed pine and laminate which is the category in which one was most likely to see someone portrayed actually cooking or cleaning. Here, then, function appears as a middle category within a dimension of identity associated with period styles (Miller 1988b). We believe it is quite common that this explicit functionalism will appear as the symbol of a class fraction that does not wish to use class position expressively, and in discussion from school choice to shopping choice would insist that their effective position within social hierarchy arises out of given needs and common sense solutions. It nevertheless forms a fairly precise niche within contemporary British society as is evident in its contrast with the other categories examined here. Of course, all shoppers use pragmatic and functional explanations for what they do, but the ethnography suggests that there are some where this is subservient to other forms of identification while there are others for whom this seems to constitute the main grounds for any identification.

In as much as class consciousness correlates with housing (which to a large degree it does) people who look down on Wood Green tends to belong to the terraced housing in the streets off Jay Road. They are householders who would identify positively with Ibis Pond. By contrast, the pragmatic shopper is best associated with people who live in the maisonettes within Jay Road, that is, people who are genuinely borderline in terms of categories which derive class from housing tenure. The pragmatic category also includes people in Sparrow Court, the more affluent of the two estates. In general people in this category would see themselves as middle class when set against the council housing of the area, but might in turn not be recognised as middle class by those who live in the other roads. My impression is that in many cases they are middle class in terms of educational background but are at a life-stage when their disposable income is often less than those who both see themselves and are seen by others as working class. Later on most of them will obtain salaries that allow them to move out of the maisonettes and into larger housing.

With respect to Wood Green their conversation tends to relate mostly to the logistics of shopping in the area. They are particularly concerned with the issue of parking, for example, whether it's worth parking at Safeway and then having to buy something there in order to get out free. They will talk at some length about this issue. There are many conversations about using buggies,

142

the problem of crowds, and the suitability of various shopping sites for children at different ages. Overall they see themselves as utilising both Wood Green and Brent Cross for their respective advantages but would not see either of them as a particular point of personal identification.

The main grounds these people give for shopping at Wood Green is that it provides a combination of cheap goods and standard high street stores. To give an example of a shopping trip to the area: a shopper went first to recycle some materials in the recycling facilities within the Safeway car-park. She then spent half an hour at the Early Learning Centre with no intention of buying anything, but simply so that her infant daughter could have a 'play-time' prior to the main shop. The rest of her shopping alternates between major chains such as W H Smith or Boots, and a series of 'cheapjacks' (see below). Just before one of these shopping trips the same shopper had bought some small birthday present items for her child at Ibis Pond, but noted that she would not dream of buying the main present there since it would be too expensive. She had waited until she was coming down to Wood Green.

Brent Cross is also discussed by these shoppers on functional grounds, and seen as one of several shopping sites to be determined pragmatically rather than according to their particular feelings about it. Discussion is with respect to a particular item that is required. Several informants also saw it as a place to go when it was raining: 'well if it was a bitterly cold rainy horrible day and I had the children I would go to Brent Cross. It would depend again on what I was getting.' Another informant noted a friend's reason for going: 'It was really raining on Sunday and I was so desperate [she said] that I took Joseph [her child] to Brent Cross. Because apparently it opens on Sundays now, and apparently it's ideal.' Apart from the advantages of Brent Cross during inclement weather, several of these informants mentioned going to Brent Cross specifically for Christmas shopping or for the January sales.

Pragmatism is also cross-cut by definitions of gender, since men regard themselves and are regarded by women as drawn to functional legitimation and also an overtly pragmatic legitimation of shopping. A wife notes of her husband:

> Brian tends to go to Brent Cross if he wants to buy clothes. But he only goes maybe twice a year and he's been last week so he won't – in the spring he might be going again. But he, like he decided before he went I need a shirt, I need a tie, I need a cardigan and a pair of trousers. And he bought this pair of trousers and he bought the same pair in a different colour. So he doesn't tend to look around that much.

Both in relation to gender and to status the idea of a common-sense functionalism comes across as a quite specific, sometimes even a quite aggressive ideology. The particular relationship of factionalism to class is discussed further in the analysis of John Lewis (below).

The sense of identity with a shopping centre

The pragmatic shopper contrasts strongly with the many informants who identify positively with the shopping centres without legitimating this identification
entirely on functional grounds. Every one of the shoppers who exhibited a
strong positive sense of identification at a personal level with Wood Green lives
as a council tenant in Jay Road. They would also all have a positive view of
themselves as working class. In some cases, particularly with pensioners, this
relationship may be based on their always having lived in the area and the
association is perhaps more with a memory of how the area once was, from
which the present is always seen as a decline. With others it is the sense of it
being 'their kind of area'. Curiously the informant who considered herself
'above Brent Cross' put it quite well when she responded to the question 'Do
any of these shopping areas have a sense of community about them?' with
the reply:

> I think Wood Green has. But then wouldn't you say that it's all a part of
> the community because everybody who's there are similar. Wouldn't you
> say that? Although they don't identify as being in a community, maybe
> they go there because they do want to be part of it.

In short, many people may not know individuals in the area but they sense
that it is a crowd composed of like-minded people. Equivalents may be found
for both Brent Cross and Ibis Pond.

Within this group, however, there are some marked differences in this relationship to the district. Firstly there are those who are frequent visitors to the
area. These in turn divide between those who shop so frequently out of necessity, and others who do so out of choice. Within the same groups there is a
division between the routine visit and the special visit. Those who visit out of
necessity tend to be the least well off and the pensioners. They do not have
cars and rely on public transport. This limits the amount of groceries they can
carry on any particular shopping expedition. One pensioner, for example, has
to carry two shopping bags, one with each hand, in order to balance himself
as he walks. At the same time, unless they use wheeled shopping bags which
are hard to carry onto public transport, pensioners have no means of carrying
more than these two bags. In most cases then they would need to go at least
twice a week.

To give an example, a pensioner living on the Lark Estate shops not
only for himself but for a neighbour who is both elderly and disabled. She
provides him with a list before he leaves. On the occasion (Miller) I accompanied him he went first to the local post office to collect his pension and pay
his rent, then we took the bus into Wood Green. He stopped first to visit the
housing offices to give a monthly contribution towards his council tax. He
then always starts by using the toilets in Shopping City since he takes water

pills. He shops at Kwiksave when it is just for himself, and at Sainsbury when going for his neighbour, but never at Safeway which he regards as too big to cope with.

A quite similar situation exists with regard to a family with four children also living on the Lark Estate and where the father is long-term unemployed due to an illness. They have more people to help carry goods but they consume a great deal more food and drink. The item that, in particular, requires these frequent visits is milk, since they consume four pints a day and feel they cannot afford the prices of either doorstep delivery or local shops. On the occasion that I went with them, they stopped first at the market section of Shopping City in order to buy gloves for their daughter who is pre-school and was with them in a buggy. They then went on to Sainsbury where they bought a variety of basic foodstuffs, including twelve pints of milk. As with the pensioner they have a strong sense of their routine in visiting the area which includes stopping off at the council offices for the payment of various bills.

These are regular shoppers out of necessity but who identify positively with the area. Within the same category but in some respects quite distinct is a third shopper, also living in the Lark Estate and who was the only individual within this group who goes to Wood Green every day except Sunday the whole year through. Dierdre was a middle-aged woman though not yet a pensioner. Dierdre lives with a partner but she certainly does not need to shop so regularly simply in order to provision the household. Rather she is quite explicit about her need to get out of her home every day and that shopping provides the basis for this. As a result of what amounts to thousands of shopping trips, her knowledge of Wood Green shopping is encyclopaedic. Dierdre has tried out every possible place where one can have a cup of tea and is clear about her favourites. When we entered a store the size of C&A she was able to view it in terms of what new stock had been placed there in the last few days. Such knowledge is obtained through a highly structured routine which provides the foundation for her trips, but with an additional element of variety to add experimentation. Most weekdays have their own specific elements that characterise the outing. For example, Tuesday stands out because this is when she meets one of her friends who works in a Wood Green store, but regularly takes a break that day to have tea and do some of her own shopping.

Brent Cross identity

A sense of positive identification with Brent Cross may arise either directly or through the view of the site as a substitute for the West End. Typically this will be found in family houses in the streets off Jay Road. One such housewife uses Brent Cross as her main shopping site for clothes and household items. But Brent Cross is still not her ideal shopping environment. When asked 'What kind of shopping do you do to just relax and enjoy yourself?', she answers:

145

The West End. That would be the more relaxing shopping. *[More than Brent Cross?]* Yes. Because they're nicer coffee shops in the West End. Well, it's lovely . . . Um, do I identify myself more with people in Brent Cross? *[Compared to Ibis Pond.]* I suppose so. I mean because Brent Cross has absolutely everything, hasn't it? So I suppose so.

For her going to Brent Cross represents an outing more than just the mundane shop:

For me if I've decided that I'm going to go on a shopping spree and I'm going to actually spend money, you know, I've got money to spend. It depends . . . Or I was in Brent Cross and have a coffee and a cake. *[And how would you dress?]* I would dress casually but I would definitely, you know, dress nicely, yes.

There are other informants who would fall within this category of people who positively identify with Brent Cross as a mark of identification rather than merely a place to shop. They were most likely to remark about the atmosphere at the centre. For example another informant noted: 'I think it's quite buzzy, even when it's empty, well it very rarely is empty, but even when it's quiet, it still feels buzzy.' While the pragmatic shopper only shops, she notes: 'I enjoy going there, I meet a friend and go to Fenwick's for lunch, it's nicer to walk round much more relaxed, so I would go more for recreation than just purely to shop. There are places you can sit down with a child and the child can have something to eat and drink.' It should be noted that actual shopping trips rarely live up to the sense of leisure implied here, but it is a strong element in conversations about shopping.

Another informant again typifies those who see Brent Cross as part of their routine shopping, though in a sense as more special than another site. She says she may not go there for six weeks but then goes three times in a week. In this case she tries to go just before opening time in order to park within John Lewis, and then on a regular basis goes to Lindys for a cup of tea prior to starting the shopping itself. She talks about one of the assistants within Lindys who treats many of the public as though they were long lost friends with effusive chats and good wishes, though in reality applying this treatment to all and sundry.

These shoppers are able to combine their sense of a visit to Brent Cross as being a bit above ordinary shopping, while usable for routine shopping. The net effect is to create a sense that routine shopping can also be special, and have in part the quality of an outing. Without Brent Cross as a discrete site within easy range, it is hard to see how this would be so easily accomplished. Given this incorporation of the site within their routines, they can also view it with the confidence of knowing their way around the centre, and its possibilities.

A sense of being below

There are no shoppers who feel that Wood Green is in any respect 'above them'. This final category is therefore limited to those who reject Brent Cross as quite unsuitable for them. All those that we have encountered to date who would be placed within this group live on the Lark Estate.

Again the rejection is most often in terms of facilities, in that they reject the idea that Brent Cross is somehow better than their own locality. One male informant noted:

> I went down to Brent Cross as my sister-in-law thought that was the place you couldn't beat it. I went down there and it cost me more to get there than I spent there, it was ridiculous. I don't see what they go on about. There's everything at Wood Green that they've got at Brent Cross except you can buy food at Wood Green and there's not many places at Brent Cross.

These assertions are, however, inevitably made in this kind of defensive tone as countering an assertion that Brent Cross is a better class of shopping centre.

The most revealing illustration of this encounter came with the same female on the Lark Estate who has just been described in terms of her daily shopping and affection for Wood Green. Dierdre had never been to Brent Cross, but as it happened when I first met her she was about to go on her first trip there. The reason was that she needed to buy something for a wedding of a friend who had a wedding list at Marks & Spencer, but the Wood Green branch, unlike the Brent Cross branch, did not deal with wedding lists. So I accompanied her on this first (and I imagine last) visit. She took along an old friend, Sandra, who had been there before, but noted immediately that 'I can't see why everybody raves about the place at all'. Dierdre's only previous experience of a larger shopping centre was at Milton Keynes, and at first her main concern was to relate everything she saw to that previous experience, by noting several times that Brent Cross was bigger. Both were a little nervous and cautious, for example, neither likes to travel on escalators, but they used each other to bolster their confidence.

Dierdre's overwhelming experience was clearly one of bewilderment. She felt that she would never be able to find her way around and was clearly disorientated. She expressed considerable surprise at the mix of shops. She would have expected more grocery shops and more places to sit and eat or have a cup of coffee. She found the proportion of fashion shops inexplicable. All of these factors made her generally agree with Sandra that this was not a place for her. Thus although she is an inveterate browser within Wood Green, she completely refused the opportunity even to look around the centre and see what shops were present. She only saw the shops in between the places she needed to visit and then left more or less as soon as she could. The sense of intimidation was certainly exacerbated in this case by her experience with the Marks & Spencer wedding list where she found almost nothing that was within her price bracket, and was flummoxed by the idea, for example, of sets of china where an individual guest would typically only be able to buy a single dish or cup. Although perhaps an extreme example, Dierdre is representative of a larger group on the Lark Estate who not only do not like Brent Cross but are quite concerned to argue the positive advantages and superiority of Wood Green over Brent Cross. There is also an awareness that their preference for Wood Green will be taken by others as confirming their low status in class terms.

To conclude this section, one cannot easily construct an autonomous classification of status differentiation from these statements about shopping centres and identity. Much of what was said might be interpreted as 'about' some other concern, such as racism or pragmatism. In the case of the final category of 'below', the evidence for a sense of inferiority is entirely based on statements that, taken at face value, would be claims to superiority. Nevertheless it is argued that the shopping centres are indeed used as points of reference through which most households construct a sense of their relative position in terms of social status. In order to do this one has to use the confidence of interpretation given by the larger ethnography in order to note the nuances and behaviours behind what is said and thereby reposition the actual words spoken.

If this use of the two shopping centres for status differentiation is then related to the objective condition of the two centres, a further problem arises. With respect to the shops themselves, in most cases there was very little to suggest that they contained connotations of social difference. Most of the larger chain stores such as Boots and W H Smith are found in both Brent Cross and Wood Green. We had anticipated that Marks & Spencer might have such status connotations but have found that apart from the food section the store does not seem to evoke a sense of class and again it is present in both centres. We believe that the majority of shops are in effect neutral. This means that the burden of difference is based upon a few key contrasts. In the next section we outline what may be the clearest example of such a contrast and one that may be particularly relevant to the use of class terms as strongly dualistic categories of opposed identity.

John Lewis as a class site within Brent Cross

In terms of positive attitudes to Brent Cross the key shop within the ethnographic survey was quite clearly John Lewis. Indeed it would be hard to exaggerate the pre-eminence of John Lewis for many of these shoppers. It should be noted, however, that many is not the same as all, and although there is a clear segment of the middle class from whom the following points derive, there are other Brent Cross shoppers for whom none of the points we are about to make would hold. One reason, however, for the prominence of positive views in the fieldwork area is that, as already noted, it is essentially middle-class shoppers who regularly use Brent Cross, since for working-class shoppers Brent Cross is too far to be regarded as a properly local shopping centre. At the same time we will argue that the centrality of John Lewis emerges where it enhances many middle-class shoppers' ideological premises about the positive elements of being middle class. It is therefore precisely when it comes to questions of identity that the significance of John Lewis becomes paramount.

For these people, John Lewis is the favoured point of entry and departure from the centre. Indeed on one visit our informant arrived on an early morning during the Christmas period when the rest of the centre opened before John Lewis, whose entrance was closed. Although a regular visitor to the centre over the years, she simply did not know where the other main entrances to the centre were located. Similarly, several of the shoppers much prefer to park within the John Lewis car park and may have a regular area that they identify with. No other car parking spot was found with which people have a particular relation.

The significance of John Lewis is a major factor in the larger relation between the shopper, their identity and the centre. For many of them the centre otherwise primarily consists of the same chain stores that they can find at other high streets such as Boots, W H Smith, the Body Shop, etc. They are not particularly interested in the smaller, more particular clothes shops. Shops such as Miss Selfridge and Top Shop are intended for a younger shopper than this group who mainly came late to mothering. It is thus the presence of John Lewis which for them marks the specificity of the Brent Cross shopping centre as distinct from a high street. Indeed they are more likely to refer to this shop than to the fact that Brent Cross is a mall as its key distinguishing feature. The meaning of John Lewis within Brent Cross emerged from a previous relationship to the larger John Lewis in the West End. The original marketing of Brent Cross was as a direct substitute for West End shopping, becoming a nearer, more convenient equivalent. West End shopping is associated by most shoppers with the traditional department store and it was the manner in which Brent Cross is constructed between two anchoring department stores that lends credibility to this claim. Brent Cross does not have Selfridges or Harrods, but it does seem to have sold itself as the 'next best thing' through this combination of large editions of the main high street stores between two department stores. Thus when asked where she mainly does her Christmas shopping, an informant replies 'John Lewis' meaning that she goes both to Brent Cross and the West End during that period.

If, however, the presence of department stores was the sole criterion then Fenwick's should occupy a similar position to John Lewis which it manifestly does not. There are four attributes to John Lewis that contribute to its special nature: John Lewis as workers' co-operative; the John Lewis price promise; the sense of comprehensive stocking that makes it suitable as a site from which to 'research' potential purchases; and John Lewis as trustworthy/sensible/helpful. Relatively few informants are aware of the structure of ownership for the

John Lewis/Waitrose group (Flanders et al. 1968). Amongst the middle-class informants, however, are a high proportion of professional, largely left-wing mothers who would see themselves as having strong moral concerns with respect to as much of their shopping as possible. One informant laid some stress on the importance to them of seeing John Lewis as a successful workers' co-operative, and felt this was a key factor in their preference for this store. Others had a much vaguer notion that there was something about the store they approved of.

Much more widely known and quoted was the John Lewis promise not to be beaten on price. Many shoppers quoted this as part of the reason they liked the shop. This cannot be taken at face value, however, since there are many other shops that make similar claims with regard to competitive price or sale value, but are regarded with general suspicion and scepticism. By contrast, the John Lewis claim was accepted as both true and given by many shoppers. But price matching has a very particular connotation here which is not the same as being particularly 'cheap'. As one informant put it: 'I like John Lewis, I always park that end so I walk through even if I don't need to go there. Everything looks nice and they have a price promise deal . . . Things aren't particularly cheap but they're good quality.'

As has often been noted the middle class often defines itself in terms of balance or compromise between opposing factors which in themselves would represent extremes. They are thus 'suburban', living in 'half-timbered', 'semi-detached' houses. The problem for the middle class is often that the term 'cheap' carries connotations of not only inexpensive but also poor quality, as in 'cheap and nasty'. John Lewis, by contrast, carries the expectation of high-quality functionalism, and therefore manages to evoke the sense of good value for money without being 'cheap' in either sense of the term. Its own brand label, Jonelle, is generally regarded as strong on quality and sufficient in terms of taste, though rarely highly stylish. In short, it may well be that John Lewis is seen as a shop that has already done a great deal of the work required by a middle-class shopper looking for what they regard as a 'sensible balance' between price, quality and taste. One presumes indeed that purchasing goods then in turn re-establishes for that person that these are their own require-ments as a shopper. It is opposed in classificatory terms to the idea of 'designer' goods or to the small boutique for clothes. Indeed its clothing is relatively dull. A *Guardian* journalist, in a witty discussion of department stores, described John Lewis as dull in the way methodism is dull, i.e. it's supposed to be, and with respect to clothing describes a 'younger woman, perhaps in her early thir-ties, examining a blouse as if contemplating her own menopause' (*Guardian Weekend* 17 June 1995). In this case, it is precisely this characteristic of its clothing section that again helps the store stand for Brent Cross in general. Although there are shops that certainly have a higher reputation than John Lewis for their clothing, the site as a whole is seen by informants as the best place to purchase non-designer goods with the implication that if one wanted the greater degree of individuality and style then one would pick a different area such as the West End or Hampstead. Brent Cross is a good place to buy clothes for work but not for the occasion of a lifetime.

The functionalist aura of John Lewis is reinforced by its strong presence in haberdashery and kitchen/bathroom equipment which makes it a kind of half-way house between furnishing and DIY stores. The extensive haberdashery and household utensil departments also make for a certain sense of the comprehensive. Another element of John Lewis, its functionalism, may appeal to the habitus inculcated through education which is instrumental in creating the modern middle class. This is based upon an explicit rationalism which utilises knowledge and research. Such shoppers may then see John Lewis as at least the best place to start an investigation for goods and prices, even if they intend to purchase goods elsewhere. Shopping may then be seen as sensible and involving minimal risk. For example, a woman contrasts her more impulsive shopping with her husband's. She notes: 'if there's an electronic thing like a video, he will buy it because he has the patience. He looks in *Which?* and all that and goes to John Lewis and finds someone.' She goes on to note that the staff at John Lewis always seem to have more of a queue of people waiting to find things out. In effect, then, shopping at John Lewis is understood as a form of research well suited to those with a background in research who can apply this to a new activity. We have certainly seen this complementarity in observed shopping. For example, a woman needing a new vacuum cleaner had looked at *Which?* but had come away with a general notion of which company products to avoid. She had not determined the precise model in the manner of a *Which?* devotee, rather having made this general conclusion she relied on the array presented at John Lewis to determine the more specific choice.

This then leads to the fourth factor, which is the idea that there is something sensible, helpful and trustworthy about John Lewis. Many shoppers who do not know of its structure of ownership nevertheless do claim to have noticed something unusual about its staff. They find them more constant, more helpful and in some ways more interested and committed to their work than is usual in large shops. This, we would argue, is in fact the case and is a direct result of the store's ownership and management. In turn this leads to the sense of a 'rational' store, where things are relatively well laid out, and that shoppers know where they are in terms of spacious layout with long-term stability in the positioning of sections of the store. This contributes to the ability of the shopper as researcher to find what they require. No one of these positive attributes can be taken in isolation; it is more that quality, sensible compromise, trust, stability and rationality are fused.

Overall we would suggest that middle-class shoppers have a strong sense of themselves as the carriers of a particularly modernist notion of common sense which looks to rational compromise and functionalism but eschews the extremes of functionalism that modern architecture or design followed through. It is in this sense that such shoppers look for goods in John Lewis but in so doing also objectify themselves and construct their own possibilities as actual or potential members of the middle class. They also find reassurance as to the values they were bequeathed by a strong strand of popular modernism that remains powerful even though it is much more rarely expressed clearly today than, for example, in the 1950s or 1960s. It is very likely, though beyond the present research, that the ability of a retail outlet to construct a sense of middle class amongst workers whose incomes would not of themselves suggest such a status is equally the case amongst the John Lewis staff as amongst their customers.

For many shoppers, no other shop can compete with John Lewis as the foundational store of Brent Cross. It would be hard to exaggerate its centrality to the idea of Brent Cross as an alternative to the West End. Fenwick's may be picked out and praised by some as better value or having a better selection, but it does not appear 'essential' in quite the same way. Indeed there is quite a common line heard from informants about Fenwick's, which is that the individual shopper praises it and sees themselves as having discovered its positive qualities. But it is as though this was a secret known only to them. While John Lewis receives general public affirmation, shoppers feel they have discovered positive qualities in Fenwick's which surprise them anew each time. Having said that, Fenwick's does not draw the almost universal approval which we have found for John Lewis amongst this particular group of shoppers.

It is interesting to note the degree to which Waitrose fails to achieve the same position. Looking at the Brent Cross layout, Waitrose might equally well have represented the conceptual end point from John Lewis, neatly encapsulating the centre as under common ownership. One can well imagine shoppers seeing this as the food shop which marks the end of their visit and thereby both complementing and completing the trip as a whole. Although many people may use Waitrose in this manner, so far at least it has not emerged as being of any particular importance to shoppers. Indeed although the observation that the area lacked food shops is often made, this is not matched by a corresponding affirmation of Waitrose as fulfilling this need. This may be partly because food shops sell in rather different ways to department stores. There is much less use of staff and Waitrose is more closely tied into a very clear structure of 'own brand' and established brand retailing that sells through the appearance of products on shelves. Also it is considered to be relatively expensive. Finally, a food supermarket, of which there are dozens in the area, simply doesn't have the same resonance with regard to justifying the visit to Brent Cross as a whole that is carried by a department store.

The cheapjack as a class site within Wood Green

Wood Green in common with Brent Cross, contains most of the major high street chain stores such as Boots, Marks & Spencer and W H Smith, so these alone cannot carry this kind of symbolic burden. For Wood Green we believe it is not a single shop but rather a particular category of shop that clearly marks out its working-class status for those who are looking to create this particular form of identity out of it. For those who work in retailing such a shop is called a 'cheapjack', though we are unsure how common this term is amongst ordinary shoppers. Our impression is that it is a term mainly known by older members of the public. There is one clear example of this type of shop within Wood Green Shopping City and there are a further six examples along the high street.

Although there is no fixed boundary between the cheapjack and general household goods stores, there are a number of characteristic features which mark this type of shop out. The mainstays of these shops are small gift items and general household goods. If there is one category of sale goods that, to us at least, seems almost essential to a proper cheapjack, it is plastic photograph frames. There will almost always be a variety of glassware, both ornamental for gifts and general utility glass. Equally common will be cheap plastic toys, general stationery and wrapping paper. In addition there are usually a variety of plastics, such as buckets and bowls. These are in some ways foundational items, but in addition almost any kinds of goods might be stocked. Remaindered books and plastic flowers are common, as are cheap cosmetics and bathroom products such as shampoo and hair styling goods. Small electrical goods and kitchen gadgets are also present in most cases. A cheapjack does not stock much in the way of clothing or food but there may be a few items of both categories, such as sweets or stockings, on display.

Apart from the range of goods, the cheapjack has a characteristic manner in which the goods are displayed. All around the shop are the most inexpensive shelf fittings together with cheap metal shelving in the centre. There are few or no notices describing general categories. There is a general tendency to have parts of the shop for particular types of goods, such as kitchen equipment, but this can be transgressed by any particular pile of goods. They are individually priced so no shelf labelling is required. Signs are usually in rough felt tip on pink or yellow card perhaps cut into a star, announcing that the prices are particularly cheap. The cashiers may be Asian, although there are also often loud cockney voices either at this point or around the shop. There may be a black male standing conspicuously near the shop entrance. The shop signs, if they have one, are titles such as 'Buy Direct', 'Quids In', 'Elite Superstores' or 'Clearance Depot'.

In ambience these shops represent something of a bridge between a market stall and a fixed shop. The placing of goods, notices, accents and the stress on cutting out the costs of higher quality infrastructure all suggest the market. In fact, however, the market area of Shopping City has no cheapjack shops; each of the shops there are specialist in particular goods, so that the hardware shop, for example, does not also stock toys and stationery in the way a cheapjack would. Rather the only example of a cheapjack in Shopping City is within the main section of the centre. This is somewhat more organized than most of the examples on the high street, but is nevertheless recognisably a member of this category. Even where such shops are actually longstanding they tend to have an air of transience. Indeed most cheapjack shops are of limited lifespan with the 'closing down bargains' not that far removed from the 'opening bargains'. They thrive on ends of leases or squatting arrangements and may be known as 'end of lease shops', though this may not actually be the case. Most people would recognise the cheapjack as the shop that fills in empty spaces in the high street. Indeed, according to the dictionary the term 'cheap-jack' was originally applied to a hawker, and one can see how in some ways this is an adaptation of a hawking and peddling tradition. But in an area such as Wood Green there are some more long-lasting examples.

This aura of transience applies not only to the shop but also to the goods, since it looks as though the shop stocks only those variety of goods which they happen to have cheap, rather than going for constant stocking arrange-ments. Transience in this context suggests that the shops are using goods which are remaindered from more expensive outlets, or stolen, or otherwise obtained from discount sources, and the savings are being passed on directly to the customer. In practice much of this is a ruse. An examination of standard brands such as Johnson's baby shampoo and Tampax suggested that they were not in general cheaper than at high street shops such as Boots, but there is the occasional item that may be dramatically cheaper. So the fact that an object is piled up as though it was about to be thrown out does not necessarily mean it will turn out to be particularly cheap. The more expensive items are on a par with the 'corner shop', and yet the cheapjack, which has attributes in common with corner shops, is in general able to distance itself completely from the idea of the corner shop as a high-priced outlet. There are some product categories such as toys or foods where prices are much lower, but the items are either poor in quality or are at the sell-by date.

It is clear that for some shoppers this is their favourite kind of shop. They enjoy browsing for bargains, and firmly believe that these are to be obtained here. For several shoppers, a visit to Wood Green is simply not complete without a browse through one of these shops, and when they have a non-specific category to buy, such as a gift, this may well be the starting point of their search. For example one shopper noted:

> Yes I always look in the bargain basements! Because at Wood Green they've got quite a few pound shops. You know pound, 50p shops. I go in there, there's quite a few good things in them shops ... I never think I must go and buy such and such. Just if I'm down Wood Green and I go in there and see something I think – Oh I'll get that because I haven't got one at home. A garlic press, something like that.

Shoppers often enjoy being inside them, feel closer to merchandise, and browse with a will. These days many shoppers are less happy with this as an accept-able context for food, owing to the high quality available within supermarkets, but the cheapjack seems much more acceptable for utility goods.

Different shoppers have their own favourites which they seem to visit on a regular basis. For example, on two shopping visits with one particular shopper she went to the same cheapjack and the Early Learning Centre but all other aspects of her shopping, and shops visited, differed between the two occasions. Since much of the rationale for shopping in Wood Green is based on its comparatively low prices, it is these shops which are thought to fulfil the promise of the area. Our impression is therefore that such shoppers almost feel an obligation to visit such shops and try to find items that would then constitute savings. In the case of this shopper, she searched for and found gift wrapping paper at a price substantially below that of such paper in most shops, but only after three visits. The cheapjack is integrated into shopping according to the type of shopper. A middle-class shopper is more likely to add this visit to their shopping at regular high street stores. More impoverished shoppers, by contrast, see the visit to the cheapjack as a part of a general routine along with Kwiksave and the market area of Shopping City.

The cheapjack and John Lewis

At first glance the cheapjack and John Lewis appear worlds apart, but we would like to argue that with respect to class as identity that is precisely the point. To understand fully the contemporary significance of the cheapjack in Wood Green the best reference point is the department store. In some ways the cheapjack is a kind of parody of the department store. It is the other shop that is characterised by bringing together a whole series of otherwise discrete categories of goods that would normally be found in different specialist shops. It has the general hardware and haberdashery items of the department store, as well as kitchen, bathroom and gift sections. It may also have a suggestion of the food and clothing departments. Just as with the department store, it is available for general browsing.

In other respects, however, the cheapjack is a systematic inversion of the values which are most clearly manifested in John Lewis. The first quality that appeals to the John Lewis shopper is its stability and sense of order. The cheapjack if anything tries to appear more transient than it actually is, and there is a manifest lack of order in the way the merchandise is presented. John Lewis is spacious and light, the cheapjack is crowded and often poorly lit. John Lewis sells on quality, the cheapjack on the basis of risk. One could never 'research' a purchase at a cheapjack in the way one might for John Lewis, because one has no idea what one will find, and there is no-one of whom to ask questions. The pleasure of John Lewis comes from its reliability, the pleasure of the cheapjack from its unreliability. Buying from a cheapjack may be a bit of a lottery, but that is a metaphor with a particularly strong appeal at present.

From the perspective of the cheapjack a store such as John Lewis may appear as a kind of haughty establishment that is trustworthy but never the source of a true bargain. The shopper in the cheapjack has a sense that, while keeping this side of the law, they are pulling a fast one over more formal commerce, by exploiting unofficial channels such as the small-scale importation into Britain of goods labelled for other European countries, or indeed buying goods that might have been stolen at some earlier stage. The atmosphere is one that encourages the sense of subverting the formal system with both cheapjack and shopper thereby extracting cheap goods at a price that would not normally be available. As we have noted, in practice such shops may often represent poor value, but just as with John Lewis it is ambience that counts. For this reason the cheapjack draws a varied response. There are certainly those who count themselves as true working class who would never step inside a cheapjack and strongly dislike them. Similarly the category of pragmatic shopper who we have noted as equally at home in Brent Cross and Wood Green would always go to John Lewis but also always go to the cheapjack since they view these as complementary forms of shopping experience.

Conclusion

The bulk of shops in Wood Green and Brent Cross have no particular class resonance. Such shopping centres are intended to be usable by virtually all sections of the community. This means that these localities may also be consumed in ways which quite ignore class affiliations. None of the evidence presented above prevents the possibility of Wood Green being consumed as a middle-class shopping site or Brent Cross as a working-class shopping site. At least one case was found of both of these opposed possibilities. One consisted of a younger shopper who used Brent Cross primarily for cheap clothes from shops such as C&A and Top Shop. The other was an interview about the use of a health food shop in Shopping City which included the following passage:

> [*Were the customers typical Wood Green shoppers or rather particular?*] No, very particular. People that you wouldn't normally see anywhere else. There was a lot of the – I don't know, Greenpeace people I suppose I'd describe them as with like the parka jackets and like the multi-locks up there. And they would actually spend a fortune on the food they actually wanted to buy.

These are major shopping sites that can give rise to a all manner of symbolic consumption. Nevertheless the ethnography brought out a clear trend that is described in the bulk of this chapter. What emerges is not only that the shopping sites may not have fixed class identities but that the same is true of the people who shop there. At the extremes there are people who have no doubts about their working- or middle-class position: they discuss the shops in terms of an affinity with 'their sort of people'. To affirm class they have a number of alternative oppositions they might employ. They could oppose the market area in Wood Green to shops such as Laura Ashley in Brent Cross, or Waitrose in Brent Cross to Kwiksave in Wood Green. John Lewis or the cheapjacks are not the sole source of such an opposition but they are one of the clearest examples of what might be termed 'shopping for class'. As such they correspond to similar patterns in other areas of popular culture, for example in the range of television programmes shown: most may well be neutral, but some soap operas and dramas appear emblematic of class as dualist values.

There remains a group for whom interior decoration, accent, clothing and opinions form a consistent grouping that could find a particular niche within the kind of 'map' of class relations constructed by Bourdieu in *Distinction* (1984). But class in the sense objectified in John Lewis and the cheapjacks is basically a relationship, and may be internalised as such. The same individual may see positive aspects of both positions and either experiment with them while regarding themselves as taking their social position from common sense rather than symbolic value (the pragmatic position) or simply switch between them as when people change accent when in different company. This may be quite transient, but there are also examples in the street of more profound changes, such as downward mobility where the older generation's middle-class aspirations have been repudiated by the younger in favour of what they see as a more 'street'-based working-class affiliation.

The majority of inhabitants of this particular street are less secure in their class position and have over the years come to consider themselves in terms of both possibilities. The opposition between John Lewis and the cheapjacks is in large measure best understood as a form of objectification by which people come to an understanding of the sets of values that they in turn label with class categories. The conclusion is similar to one drawn from a study of supermarkets in Trinidad (Miller 1997) where class position is far more fragile as a result of the recent oil boom and recession and where the same individuals exhibit markedly opposed behaviour in different supermarkets which become, in effect, practice grounds for the possibility of a more stable class opposition. In seems that in London, too, as what were once more stable class positions based in clear occupational expectations are in decline, then elements of consumption such as shopping increase in importance as central instruments within a continual process of class construction.

It would be quite wrong to see this as some kind of new postmodern ambivalence. Schama (1987) makes a very similar argument with respect to seventeenth-century Amsterdam, as is made here for contemporary London (and elsewhere for Trinidad). In all three cases there is a stronger case to be made which relates such findings to a more established literature on contradiction within comparative modernity (Miller 1994: 58–81). The problem of job insecurity combined with better possibilities of class switching is not somehow less 'authentic' or more superficial than a consistent identification with a particular cluster of values. Most of us watch television and participate in media that celebrate both these 'class' positions and we can increasingly envisage ourselves in relation to both. Many people inhabit not one or other site, but encompass the relationship itself and the field of difference. This is not at all the same as classlessness, nor some kind of inauthentic play. Such an emphasis is important, however, in helping us to understand how and why institutions such as shopping centres might be playing an increasingly active role in the construction of identity.

8

ENGLISHNESS AND OTHER
ETHNICITIES

This chapter represents a conclusion rather than a premise of the project in as much as we had hoped not to presume 'ethnicity' as a key parameter in the relationship of shopping and identity, or to isolate it as a variable. There is a literature which employs the concept of ethnicity in order to analyse the different relationships of those designated as ethnic groups, most often to business but also increasingly to consumption (e.g. Venkatesh 1995) but we had no prior intention of following along this path. The most interesting trajectory in the current academic study of ethnicity in Britain is one which focuses upon the new identities that are emerging, most especially in relation to music and which often transcend any simple classification by origin (e.g. Back 1996; Gilroy 1987, 1993a, 1993b: 120–45; S. Hall 1992a). But, while not wishing to presume or impose ethnicity as a perspective, it would be equally reprehensible to suppress something that emerges clearly from the field studies themselves. It soon became evident that for many people terms such as Cypriot, Asian and Jewish were of importance not simply as background classifications evoked by relevant questions, but were categories which many people expected would and should stand for particular styles of interaction with the two shopping centres under study. In short, ethnicity emerged as part of the identity of the sites themselves, and therefore an aspect of the overall relationship which we would have to highlight.

As has already been noted, one aim of this study was that each method should pick up its leads and sites of enquiry from those that preceded it (while also being prepared to challenge such previous findings). Both the questionnaire and initial observations of our field-sites suggested that specific groups might have particular relationships to the centres with which they identified. The selection of Cypriot youths for a focus group followed from clear evidence already emerging that this particular community was known to use Wood Green as a leisure site for 'hanging around'. Similarly the selection of the Jewish women's group (which came very late in the study) followed not only the statistics on the use of Brent Cross by the Jewish population but also the sense of the importance of this relationship that was emerging from the ethnography. The only exception to this logic of enquiry with regard to the focus groups was the West African trading group. This group was recruited by Vicky Knight, a graduate student at University College London, working under the supervision of Rowlands, who had an independent interest in these questions.

But the 'logic', based only on the degree to which a community was present, would anyway be rather limited. It implies that one studies only those groups where a strong relationship to the centres has been found. The crux of this chapter, however, is an argument that such strong identifications can best be understood only through a series of contrasts. It is equally important to understand those groups who have a major presence in the area but who don't seem to identify with the centres. Furthermore the term 'identification' covers also the process by which one group identifies a site with another group even when the latter does not itself make any self-identification with this site. Finally the concept of ethnicity as a marked group makes sense only with respect to another, usually dominant, 'unmarked' and in a sense un-ethnicised group understood as British (or, more accurately, English) against which the specificity of ethnicity is defined. A fuller understanding of the relationship between a site and any one community depends on a much richer context than could ever emerge merely from the study of that single group in itself. This chapter will therefore concentrate on the contrast between the various communities selected for study.

As in other chapters (such as that on class) the concentration upon shopping centres of itself shifts the implications of focusing upon a particular social category. Our primary interest is in a process of objectification through which identity is not treated as a pre-given attribute of persons or groups whose relationship with some external form or institution is then studied. Rather, the materiality and spatiality of the centres is understood to have the potential to become itself one of the media through which that sense of ethnicity is derived. For example one might have the 'Wood Green' Cypriot or the 'Brent Cross' Jew as coming to signify within and outside the respective communities a particular type or stereotype that is used to differentiate sectors or sometimes to become symbolically dominant for a much wider group. Within London the significance of location is very evident (cf. Keith 1993 on the 'symbolic geographies' of Brixton, Stoke Newington and Notting Hill, all defined as key sites in relations between black people and the police). Brixton may stand for Black, and Southall for Asian, Kilburn for Irish, Golders Green for Jewish in a manner that constrains the possibility of self-expression even of those who have never been to the places in question. An individual born as a Jew may not wish to be associated with Golders Green, which they may never have visited, or if they are entirely secular with being Jewish. As in all such attributes, these then become contested elements which can be seen positively, negatively, with ambivalence, with irony and in various framed and contradictory points of reference.

There have been numerous recent studies which emphasise a relational view of ethnicity and which attempt to interpret cultural intermixture (creolisation) in a positive light. Earlier (racist) views interpreted any 'dilution' of the 'purity' of one group by intermixture with another group in unambiguously negative terms. Far from leading to 'degeneration', as some nineteenth-century scholars argued (Young 1995), recent debates have attempted to define such 'hybridity' in more positive terms, involving a creative blending of different cultural traditions. Back's (1996) work on youth cultures in South London suggests that this is true of musical forms such as *bhangra* (the blending of Punjabi folk and African–American music) and the work of artists such as Apache Indian. Likewise, among the Punjabi community of Southall in West London, Gillespie (1995) suggests that the consumption of even the most global products such as Coca Cola and McDonalds have highly localised meanings depending on the mediating influence of different cultural traditions (such as parental restrictions on children's eating habits and the attractions of specifically American constructions of the 'teenager').

By focusing on the material objectification of ethnicity we are able to loosen the normal emphasis on personification, that is ethnicity as an attribute of people. One of the principal conclusions of this chapter will be to dissolve the single concept of ethnicity as a generalisable category. Rather, as will be shown through contrasting five different examples, the term ethnicity obscures some extremely diverse forms of community and spatial identification each of which has quite different implications for the process by which locality and identity are articulated. The problem would be to start with a thing called 'ethnic identity' and try to understand shopping activity from that starting point. Unfortunately this remains much too often the basis for discussions of identity. Although the concept of race is generally discredited, the popular and much-employed concepts of ethnicity and, increasingly, 'culture' occupy a position so close to that once occupied by race as to be virtually indistinguishable.

The language of 'race' and 'ethnicity' is extremely vexed. While 'race' (and its associated oppositions between Black and White) was once rejected for its crudely biological overtones, it has since been reclaimed as a social construction with specifically political meanings (cf. Gates 1986). The slippage from 'race' to 'ethnicity' has itself been condemned as moving too quickly to a cultural realm (Lawrence 1992), while more recently the specificities of 'Black' and 'Asian' experience have been distinguished (Modood 1988), together with an insistence that the majority-society's 'Whiteness' should also be recognised as a racialised construction (cf. Frankenberg 1993; C. Hall 1992). With a recognition that racism may be taking an increasingly cultural form (Blaut 1992) there has been a re-emphasis on various 'new ethnicities' (S. Hall 1992a), acknowledging the politically and culturally constructed nature of social difference, with no guarantees in nature.

The diversity that will become evident from the case studies to be presented may be ranged along a particular dimension, which simplifies but also clarifies their differences. At one end of the spectrum lie groups who have not been entered into mainstream discussion of ethnic stereotypes and are themselves relatively recent immigrants who may retain strong endogamous connections. The prime example here would be African traders, but there are many potential alternative examples ranging from East Europeans to immigrants from various South American regions that would make the point equally well. Here the concept of 'culture' as used colloquially is probably most appropriate in implying a normative set of dispositions arising mainly from their connected history and retained networking within the particular community.

The other end of this dimension is constituted by groups who are central to generalised discussions and constructions of ethnic stereotypes, and whose identity in Britain is largely constructed in relation to these widely held stereotypes that are often the subject of explicit discussion both within and outside the communities in question. The clearest examples would be the West Indian community and the Asian community. In both cases these are communities that are extremely heterogeneous in origin, a Jamaican of African roots may have little background in common with a Guyanese of Indian roots. But within the British context they are brought together as all sections have opinions about 'Blacks'. In many cases they (especially the younger British-born element) may indeed identify with the notion of being Black and with both the positive and negative associations that come to bear upon being Black and British. A similar case would operate with the combining of groups such as Sylhetis and Gujeratis as 'Asian' or 'Paki'. In these cases it is impossible to consider identity outside of racism as a primary point of reference.

Quite distinct from either of these cases is the concept of 'Englishness'. As will be argued in more detail later on, this may be the most important ethnicity in the process of creation at the present time and is one where again people are starting to reconstruct themselves around this concept largely in view of the values and associations, both negative and positive, that are being currently concretised within it. While there is much less explicit debate about this category, as an implicit attribute of both people and places (as in the assumed Englishness of a shopping centre that is located in London), it has a central position in the wider discussion (cf. Taylor 1991).

Between these two extremes are a number of categories such as Cypriot and Jewish which represent mediated points, in that compared to the West Africans there is less clarity about a common 'culture' with which people would necessarily identify on the one hand, and also less current concern to use these categories within the larger debates over value and identity. In the case of the Cypriot community this is still an identity in initial formation, in the case of the Jews it represents a decline in the use by the 'host society' of this particular ethnicity as a point of objectification compared to the 1930s when Jews would have occupied the centre stage now occupied by Blacks and Asians in discussion of the generic category of immigrants.

To conclude: there are two key variables which may or may not be related in any given case. The first is the actual homogeneity and similarity of 'habitus' (Bourdieu 1968) within a given community, and the other is the degree to which an abstract discourse has been created within the host society pertaining to that group. Either or both of these may use a particular space or place as a medium of objectification. It will be shown that these two factors make a tremendous difference to the way in which locality and identity will be articulated. The argument of this chapter is summarised in Figure 8.1.

Decreasing identification   Decreasing identification

West African traders	Cypriot youths	Jewish women	'Black'
Example of recent immigrants whose identity remains closely associated with place of origin	Example of group where local aspects of identity are currently being constructed	Example of group where fear is growing about possible loss of its own sense of distinction	Example of group where identification with locality is made by others and projected onto this group
Little evidence of identification with shopping centres	Strong identification found with Wood Green	Strong identification found with Brent Cross	Strong identification made by others of this group with Wood Green

Figure 8.1 Ethnic identification with Brent Cross and Wood Green

West African shopping

The material for this section comes from two focus group discussions together with a pilot project in advance of a full ethnography by Vicky Knight. Although Knight's discussions include West Africans who have now permanently settled in Britain and are becoming part of the British 'Black' population, the main emphasis of her work is on Ghanaian women traders for whom London is part of a larger network of sites whose orientation remains within West Africa. As such this community exemplifies a particular kind of diasporic community who primary unit of location is transnational rather than national. Such communities have existed throughout recorded history (see Ghosh 1992 for an exemplary study), but it is likely that modern transportation has made them a far more viable and common phenomenon throughout the world. In such cases the diasporic community is itself central to the main economic activity which is trade based on retaining close kin and community contacts across a wide range of North American and European countries linked back to the homeland through constant communication and frequent visiting.

In general the focus of these traders as shoppers is on sites which are useful for their own particular and largely pragmatic purposes. These are mainly either markets which are important as sources for the items they trade in, or shops which sell the particular West African foodstuffs and other goods which they use in daily life. The retention of a strong sense of homeland and tradition is of considerable importance and they will spend much time and money in travelling to shops that specialise in West African items such as foodstuffs and clothing. They may continue to remit funds back to Ghana and invest both financially and culturally in their natal regions.

For such a community, although they may represent a major presence in sites such as Wood Green, the shops and their localities do not become major points of identification. Their identity is rather invested within sites back in West Africa. This is evident in their discussion of shopping for Christmas. They are very little involved in buying Christmas presents for people here. By contrast this is clearly an important occasion for sending things back home:

> What I do is I send very much quality things back home. At Christmas time I send a lot of things back home.

> All Ghanaians do it. No Ghanaian will let Christmas by without sending anything to their relations.

Furthermore the things they note for sending home are traditional West African items such as traditional cloth, 'Dutch wax', sent for a mother to look well dressed at Christmas, rather than English items. In short, the ethos is of a group that will 'send coals to Newcastle' rather than take advantage of their immediate access to what back in West Africa might seem to be exotic English produce.

As such these traders are primarily concerned with shops and markets in as much as they serve particular economic and social purposes during the time they are living here. They also assume that their own ways of shopping and bargaining are West African and are a point of contrast with the behaviour of English shoppers. For example, they note the lack of queuing in African shops, or that they merely pick the things they need, without the browsing and choosing on site that they see as characteristic of English shopping; so the style of shopping as well as the goods they buy may be used to reaffirm the sense of difference from the host community. Their problems are more likely to be the desire to obtain certain goods at a low price which they intend to sell abroad for profit, or the difficulty and expense of obtaining supplies for their style of cooking. They complain that this may not later be appreciated since from the point of view of the families for whom they cook the results are much the same as they have ever been. Their lack of local assimilation is exemplified in the reply to the question as to whether they ever eat pizza. Most said they had never tried it, and one claimed to eat it 'well maybe once a year'.

For such groups, as long as their orientation remains focused upon a site abroad, local sites have little implication for identity. Rather their identity is mainly affected by their sense of difference from the local population. It is only when they become immigrant as against merely diasporic groups that this situation changes. The change can, however, occur quite swiftly. Within one of

164

the focus group discussions there is evidence of individuals who may be in transition between the status of being West African to being Black British. In that case their concerns and responses become much closer to that of the Black population in general. In this focus group the conversation is about 'Black' people as much as West Africans, and the cultural resources may include 'deconstruction' as learnt from a media studies degree, a streetwise knowledge of the specifics of particular retail chains, and also a sensitivity to racism. They are also marked by a relative reluctance to send presents back to Ghana. Their use of ethnically differentiated goods such as traditional clothing is starting to become restricted to particular events such as life-cycle celebrations, or attendance at church.

Within the ethnography many similar cases could be found. Sometimes the violence which had precipitated the migration to London from areas such as Somalia and Kurdistan did not permit the same continuity of links to the place of birth, but the orientation remained to a community in exile rather than to local sites. The major changes that can be observed reflect those of previous communities of migrants, such as anxiety over the loss of identification with the homeland by their children, and anxiety over choices of school in London. Although some indication of identification with local shopping is emerging, it is clearly secondary to a much more important concern with education.

Jews and Brent Cross

The material used here on Jewish identity with Brent Cross is taken entirely from a single focus group. Within the ethnography there were a number of households within which one partner was Jewish but only a single household where both partners were Jewish. The question of Jewish identity came up in a number of cases in relation to Christmas shopping but was not specific to the shopping centres themselves. What did emerge was a strong positive identification with Brent Cross amongst the middle-class segment, but this was not articulated directly in terms of Jewish identity within the ethnographic context. By contrast, the focus group was based explicitly upon an already existing Jewish women's group, whose intended subject of discussion was their relationship to Brent Cross. By definition then these were women who have a clear identity as Jews and as Brent Cross shoppers. It is likely that there are other segments of the Jewish population who would by no means share or feel comfortable with the kinds of identification made by this particular group, but most would recognise that within the London Jewish community there exists an element for whom Brent Cross has become a highly significant location.

What comes over throughout the discussion is a concern on the one hand to define what it is to be Jewish in relation to shopping, but at the same time to use self-deprecating humour to keep a distance from these same identifiable features. Indeed the dominant characteristic of this lengthy discussion is that it includes a constant and largely successful striving to be funny, which in turn comes across as an effective instrument in both expressing and living with this sense of ambivalence.

The identification with Brent Cross emerges early on, in a sense of guilt as to the degree of familiarity it has for the children:

> So he's obviously been brainwashed by me as a child – Marks & Spencer's equals sandwich, equals fountain. It's quite sad but the kids used to like it there. *[At this point the other members of the group signified they empathy with this degree of identification by taking up the chant 'Salmon sandwich, salmon sandwich' in the background.]*

A more specific relation to the community is evident in the statement:

> you know, you know, the person third row along is going to have the same outfit which is a worry to some people. *[The third row here refers to the sitting together in the ladies' gallery of a synagogue.]*

This sense of over-identification reaches its apogee in an anecdote about a lunchtime quiz held by a Jewish group:

> it was a quiz and the marathon round which was the big round that happens while you're eating your – well it wasn't very much lunch because everybody was slimming, there were like a hundred women that were all slimming – the marathon round, instead of identifying you know filling in the tube map, the underground station or whatever, it was actually a floor plan of Brent Cross, both floors, and you had to fill the shops in and I think there were 86 shops in Brent Cross and one table of ten women filled every single shop in *['I can do that', 'Yeah right']* They filled every single shop in and they could even say . . . and they actually knew which shops had closed.

At the same time that they note this degree of identification, they also see this as a negative attribution of the place. The implication being that Brent Cross is itself degraded by such an overt association. Three examples make this point:

> It's those words that sort of . . .
> I'm going to Brent Cross.
> It's a dreadful name actually.
> But only 'cos it's got the Jewish connotations.

> I don't want, I don't want to be tarred with a brush or – I said that all
>   wrong but anyway I don't want to be –
> Labelled?
> Labelled – that's the word – a Brent Cross shopper.

Why?

Because of the stigma I feel attached to the women that get dolled up
and go into Brent Cross just to be there.

I don't think there's anything particularly, it's just a stigma attached to it,
but it's in a Jewish area and a lot of Jewish people go there and being
Jewish we notice other Jewish people.

It should be noted that in each case such statements were met by rejoinders
both in pointing out that other people use Brent Cross such as Asians and
middle-class non-Jewish English but also that the Jews in question were not
really the kind that are being objected to. This establishes a basic framework
of ambivalence, a fascination with the identification of being a Brent Cross Jew
but also a horror of it. From here the relationship can be explored through
more complex and particular motifs.

First there is the identification of the particular kind of Jewish woman that
is seen as the point from which informants could create their sense of distance
or atypicality. With reference to the lunch quiz the informant noted:

But these I suppose were the typical, the sort of ladies that wear glitter
and wander round there and lunch and shoulder pads. But then Julia
who doesn't wear glitter, lunch or shoulder pads could fill the shops in
– well she has lunch but not . . .

She eats more than a lettuce leaf. I promise you Julia does not have lettuce
for lunch unless it's got coriander and limejuice and an anchovy.

And serve it in half a red pepper – sorry. But I couldn't believe that and
that was like this, it was everybody's worst nightmare of stereotypical
Jews.

A shorthand for women who are over-identified with the site in one conversation revolves around the concept of rhinestone as in:

But don't you think that perhaps the rhinestone women, if we can label
them, OK they are there but they just stand out more and we think that
everybody's like that. There's more women with buggies than there are
with rhinestones and there's more women with lots of children.

or

Fenwicks target the rhinestones.

An alternative attribute came in the form:

You have to have your shoulder pads surgically implanted as well.

167

When asked by the convenor (Holbrook) to specify what was 'Jewish' about their shopping, the discussants found it harder to come up with examples. The most concrete was the need to buy a new outfit for the New Year synagogue attendance and the importance of hats worn by married women at synagogue as in:

> I don't know who else buys hats other than Jewish women.
> But hats is very middle class obviously, but so is you know the fact that the Jewish holidays, that people buy new clothes for, or I mean possibly you probably need perhaps a certain kind of smarter clothes.

Even this is contradicted by others pointing out that most Jewish shoppers are no longer religious and rarely attend synagogue. This suggests that the identification is essentially a social phenomenon rather than anything truly distinctive, thus contrasting greatly with the West African case. This may suggest that locality has increased in importance as alternative forms of difference are now highly attenuated. The discussion reveals a fascination for recognising other Jews within the Brent Cross context, preferably through overheard conversations or gestures and a kind of collective admission of identity through association with the site itself.

Brent Cross is not left as a unitary phenomenon in this regard; particular sites are noted as important, for example the fountain where people have sandwiches. Individual shops are also discussed in terms of their relationship to Jews. Stories that suggest anti-Semitism are shared, as in:

> I have quite a lot of contact with . . . and in fact as I was saying to Julie he [specified staff member] came in to see me today, and the collection that I was selling is for next summer, and his attitude towards the women who purchase from his store is purely anti-Semitic, and it's actually quite horrific to hear what he has to say.
> Well send him all the wrong stock then.

Or

> . . . used to have a racist policy.
> My mum was refused a job in there and she was told, this was years ago, she was told because she was Jewish she wouldn't get a job there.
> My father won't go . . . didn't used to employ anybody other than White Anglo-Saxon Protestants basically, they had a no Black, no Jew policy when they could for years.

These women related to the business of retailing that lies behind Brent Cross much more explicitly than most shoppers, whether this was to salespersons or retail technique. Many Jewish households have experience in the clothing trade and this is the dominant retailing component within Brent Cross. They are therefore able to switch from an insight in the supply of shops to identification with selling through to their role as consumers quite easily. It was otherwise very unusual for middle-class shoppers to identify with any role other than that of shopper.

The focus group discussion contains many such examples and insights which are testament to the high level of identification with Brent Cross. It appears that Brent Cross is central to the changing sense of what it is to be Jewish for these women. It is precisely their difficulty combined with their obsession to define themselves as a community that is at stake. The links to any genuinely distinctive 'habitus' have reduced to a few concerns with synagogue such as hats and New Year outfits. It is more evident for the extreme 'rhinestone' characters, but the discussants distance themselves from that stereotype. In the absence of alternative distinctions Brent Cross becomes an important site within which they attempt to identify and define their own community. The leitmotif of the discussion consists of questions about whether it is still possible to recognise and affirm solidarity with other Jewish women in this context. The ideal of recognition is based on the model of a family which despite different life courses come together and are met again within this shopping context. Two conversations illustrate this point:

> You know them or you think you know them.
> They look like your family.
> They're family exactly . . .
> Though we don't look the same.
> We all look different.
> But we're all unite – we're all one family.
> You know if I saw you in Brent Cross.
> I don't know, you just know, there's a link.

> It's a small world but it's a very small world if you're Jewish.
> It's frightening, I mean you never know.
> You can't. You can lose touch with people who aren't Jewish like people you went to college with but somehow if they're Jewish you can always trace them somehow.
> I think it comes from, there's some funny things, there was one girl that I was at college with that I didn't know at the time but are now very good friends and we did this Jewish geography afterwards and we've remained friends – now there is a bond there.
> You always feel being Jewish that you belong to an extended family, it is there. You will be looked after wherever you go if you say, I'm stuck and I'm Jewish, you will be looked after.

The importance of locality within the Jewish diasporic community had already been established through the laws that forbid driving on the Sabbath. This meant that groups had to live within walking distance of each other in order to pray. This traditional use of locality is here transferred to a new site which relates to another 'secular' tradition of the British Jewish population, namely their involvement in the clothing and retail industries. The point by the end of the conversation is not that Brent Cross is Jewish, but that there is a Jewish appropriation of Brent Cross which is distinct from all other appropriations and which therefore comes back as a mirror image confirming the identity of a section of the Jewish population. This is consistent with a general ambivalence about all those behaviours which have been used to create this sense of identity. This ambivalence in turn generates the constant joking about 'over-the-top' examples of Jewish identifications with the site that dominates the discussion. In this sense it is the humour itself which provides the best evidence for the particular nature of this relationship.

Cypriot youths and Wood Green

The decision to interview a group of Cypriot youths followed from early evidence that they are the main group that use Shopping City for hanging around. They are not therefore a representative group 'standing for' ethnic minorities, but were selected for further enquiry as a result of evidence that showed them to be unique in this regard. This is quickly confirmed by the focus group and as with the Jewish women there is also an immediate distancing of those interviewed from too direct an association with this activity. This is evident in the following edited form of the early part of the conversation:

> There's a group of girls and boys usually Greek and Turkish that's just like the majority of the ethnic – you know the majority. And they all congregate outside Mr Byrite's . . . They would like stare at people and take the mick out of people and it made people feel really uncomfortable . . . No one will walk through into the shopping city through the main doors they'd have to walk through Boots because it makes them feel so uncomfortable . . . There's seats and walls. So it's not like they've chosen one shop to just congregate outside of, there are walls and seats that they're probably you know starting to sit down and meet their friends all sat down and they've said 'oh I'll meet you outside Mr Byrite' and they all sit there. It is so uncomfortable even for me like I go down on a Saturday and go past it's just horrible . . . You might go past then there's sort of like 40–50 people just big crowds . . . You don't necessarily have to be in that group but if you're seen anywhere even in the shopping you're called the 'Byriters' and that's – you just get the name . . . *[What do they do to make you feel uncomfortable?]* They stare 'cos they take the mick out of you – They stare at you – If you're not one of them you know – Not only that they've got a bad reputation so you want to keep your reputation good and if you're seen with them – especially when they're Greek.

Those who hang around are mainly youths, and some of the discussants are embarrassed about a section of those who hang around and who are clearly a good bit older than they are. Although this group is identified ethnically, in many ways this is cross-cut by age. This in turn may be related to the identification of the youth groups of other minorities. At no point in the conversation is there any sense of antagonism against any other minority group, rather there are several statements of sympathy with black youths (see below). The association seems to be with the Wood Green area itself as in the following exchange:

> [Do you think there's a race problem in Wood Green?]
> Not in Wood Green.
> Not Wood Green 'cos it's like . . .
> It's my culture ain't it.
> Yeah, it's so multi, it's definitely not in Wood Green.
> It's like half the staff come from Wood Green anyway and then multicultural as well.

By the same token these youths do not feel at all comfortable at Brent Cross, where the sense of alienation does not come from ethnicity but from their status as youths. As one of them notes:

> Yeah Brent Cross if you go into the expensive shops in Brent Cross you see all the sales girls going and they start looking at you like that because they don't know how much money I've got. Just because I don't dress flash and you know or I might go straight after work and I look a mess. So and but they sort of like keeping like their eye on you.
> They're very reluctant to serve you, they prefer to serve very posh looking persons than to serve you, they might not spend anything.
> And the layout of it like in the cross and that big waterfall in the middle and at Christmas when they've got that Christmas tree just makes it look nice. And if you walk into Wood Green it's not that, you know, it's just normal you don't feel impressed by Wood Green.

The youths would like to have an unequivocally positive association with Shopping City and clearly feel this is spoilt by the negative attributes of the Mr Byrite crowd, but the blame for this is laid on the external environment. Partly this is the sense of harassment related to the security guards who are criticised at several points, but it also is seen as a result of having little else to do, which they feel could be rectified as in the following conversation:

> The only criticism I have is that they should have a lot more going on in the actual shopping city 'cos a lot depends on attraction, and attraction yeah that's –

It's true actually 'cos I saw this dancing once and everyone was crowding
round.
Try to organise cultural activities . . . in the Shopping City?
We had a fashion show there as well which was really good it really got
the crowds and everything, there was a Greek radio show there.

As with the Jewish group there is little that can be specified as 'ethnic shopping'. Although there is one shop that specialises in Greek food within the
food hall, there is actually an abundance of availability of such foods all through
this area and Haringay more generally. Quite apart from Cypriot-owned shops,
Asian shops tend also to stock these goods. The supermarkets also stock 'Greek'
food items but the youths are quite clear that the latter are regarded as inauthentic versions intended mainly for the non-Greek population. In any case,
there is less an association with shopping as an activity than is the case with
the Jewish women since here it is the youths' parents who would do most
of the actual shopping. The sense of identification comes mainly from locality
as a place to meet and hang around.

The tightest social groups are based around schools as one school age informant notes in conversation:

[So what about Wood Green Shopping City when you go there, I mean how
often would you say hello to somebody?]
Every five minutes.
Every other person. The whole school's down there.
It's school is it?
Yeah school mates, my school is in Southgate and people from Southgate
will go down to Wood Green.

The evidence of the ethnographic study suggests that older Cypriot housewives do not have the same kind of association with Shopping City as some
Jewish housewives have with Brent Cross. The five relevant persons (four women
of Cypriot origin and one male Greek student) may be a rather small group
from which to generalise, but a consistent attitude emerged within the ethnography. Although the Asian community was probably the most resistant to being
interviewed, this group was the only group that was happy to be interviewed
but was highly reluctant to be involved in accompanied shopping, which therefore did not take place at all.

One possible explanation for this (for the women involved) could be a
reluctance to be seen in public with a male ethnographer. That it may
have been more than this, however, emerged in a more general sense of
ambivalence about identity. Compared to the Jewish women who were keen
to establish their specific identity, the four Cypriot women tended to play down
any sense of ethnic identity. During the course of interviews two distanced
themselves from specific association with Greek food, and one notes how

172

few Greek decorations she has, another refers at one point to the 'foreigners, Turkish, Kurdish' as opposed to the English. One states her preference for Brent Cross, another prefers Wood Green but only because she can't afford Brent Cross and the latter is less suitable for sitting down for a coffee. All of this suggested little or no positive identification at this age for any affinity between being Cypriot and Wood Green, or at least this is not made explicit.

Why should the association be with a particular age group rather that with the Cypriots in general? Underlying this may well be an ambivalence about Wood Green based on the dynamics of inter-generational tension. The evidence is slight, but it suggested that there exists firstly a more elderly segment of the population which positively identifies with Greek culture and is most fully expressed as the tight knit community which may be found at local day-care centres. These would be the grandparents of the youths interviewed. In between comes the generation of the parents encountered through the ethnography. These emerge as much more ambivalent, influenced by a desire for integration expressed in a distancing from things Cypriot and also concerned with the future of the next generation. The youths seem, by contrast, to have moved from an identity based on simply cultural continuity as experienced by the elderly and replaced this with the generic sense of Greeks and Turks as a constitutive part of the 'ethnic minority' scene being constructed within the British context.

It is, then, the parents whose anxieties come to the fore and may be expressed even in their reluctance to shop with an ethnographer. One of the mothers discussed her own history in terms of the desire of the youth to break away from a sense of claustrophobic tradition:

> I don't know if you've got any Greek friends or anything but what happens is when you've got a daughter or whatever you go out looking for a groom and you get people coming round your house looking at you like you're a piece of meat, you know what I mean, and it used to really piss me off, and that and I went against my mum's wishes and I just married the first bloke that I found practically. You know he was a good looker and what have you and I thought sod that, and so we did and he was Greek and I got pregnant and had my son just before I was married actually had my son, so there is no way my parents would have agreed for me to marry. You know he wasn't in our class, you know they looked at classes and that and I married him and despite their . . . I spited myself in the end I realised, because you know we had nothing really in common because he was from a poor background and he had nothing to really offer me. I was used, and in those days money didn't really bother me because I was used to having.

This inter-generational conflict over identity comes out within the focus group as a general desire by the youths not to be seen within the shopping centre with their parents. As one put it:

> I feel really bad sometimes because mum will go 'do you want to go to Wood Green, I need to get some stuff and you need a new pair of shoes?' and it's like 'oh I can't today I need to do this', 'alright then next week?', 'yeah OK' and I quickly make arrangements for next week. I feel bad but I feel embarrassed being seen with my mum.

The evidence suggests a strong sense of identification with Wood Green, but while in the Jewish case this pertained to housewives, in the Cypriot case this relates specifically to youths. The reason for this difference lies in their respective stages of immigration. If one reads accounts of Jews in the East End early in the century there is a great deal that is parallel to that of the Cypriots today. There were large groups of youths, often rather more aggressive towards both the police and outside communities, but also involved, one suspects, in constant quarrelling within the generations of the families themselves as they expressed their secularisation from tradition.

The movement from ethnicity as a sense of continuity to ethnicity as something being constructed through the use of shopping centres as locality is expressed in a generational distinction between the youths and their parents. The fact that it is also based largely on hanging around rather than shopping *per se* makes relatively little difference to the possibility of incorporating a shopping centre as a defining attribute of the group. What both the Jewish and the Cypriot focus groups suggest is the possibilities of using locality to create particular facets of the larger category of identity. This is not particularly specific to religion or origin. The finding is similar to that of Mort's observations of the way the gay community in London developed a particular topography of Soho gay life in the 1990s (Mort 1996: 149–99).

The discourse of Blackness

The concept of 'Black' provides the opposite polarity from that of the West African with which the chapter started. This label does not refer to any actual 'habitus' or cultural tradition of those designated by it, rather it relates to an explicit discourse about 'Blacks' which is found distributed throughout the ethnographic and focus group material. The concept of Black seems an extremely apt example of what psychologists who used Kelly's (1955) personal construct theory called cognitive dissonance, that is an almost complete absence of fit between empirical experience and the retention of a concept that purports to relate to that experience. This is evident from both Black and non-Black informants. There were more West Indians than any other minority group within the ethnography, but none of them was easily categorised as Black. Many of them were elderly and came from West Indian traditions where the concept of Black is largely absent. Rather they are defined by place of origin as Guyanese, Jamaican and Trinidadian and would thereby regard each other as extremely different. The younger West Indians were female which again brings up this dissonance, since the term 'Black' that emerges in the discourse about shopping is largely male. The Black that emerges from the discourse of others is violent and threatening while the female informants tend to be involved in areas such as nursing which is the antithesis of these attributes. Such is the dissonance that it is not improbable to find a West Indian informant who is as anti-'Black' as any other informant. Few shoppers would not associate themselves with a community discourse defined by its opposition to notions of violence and evil that have become attributed to the generic concept of a Black threat to the shopper.

This should not be exaggerated; there are many elderly Whites on the estate whose concept of Black is rather broader and is directed as much against the female nurses as male muggers, to become a general racism against all people with dark skins. There is no dissonance in their case since their racism is actually directed both as discourse and to the actual Black individuals they encounter. Equally there are many people who are able to frame this discourse of Blackness within a category of racism which they find abhorrent, such that they are able to view sites such as Wood Green without this being refracted through the lens of such discourses. In short when dealing with this diverse category of 'Black' we have to confront a number of different phenomena some of which bear little or no relationship to any actual persons who would identify themselves as Black.

The most salient concept of Black is that which has been described with a number of illustrative quotations in the previous chapter on the family. It is used by some of the non-Black population to characterise their sense of fear and threat within Wood Green. In most cases the association is negative, but two quotations from the Cypriot youth group are instructive, partly because they represent a positive identification with Black as victims of racism and also because they provide observed evidence of the way actual Black people are objectified as the embodiment of racist concepts by groups such as shopkeepers and the police:

> I work in a shop and I don't really want to name the name of the shop because when I'm working there and a group of young Black girls come in or something, I get told to stand and watch them and follow them around. And I feel so uncomfortable because I'm not racist, I don't like to you know stare at them and I think to myself people do it to me and I don't like it, how do you think they're going to feel because they're Black as well and they always get it. But I'm always told to watch them.

> This might be going off the subject a bit but what I've seen in Wood Green is with the police as well, they go down to Wood Green there's groups of White boys together when they come across two Black boys together they'll stop them and search them in the middle of the street. And it really intimidates them in front of everyone being searched and a

young Black boy once they searched him and he was saying 'what are you searching me for there's nothing on me?', he'd just come over from Nigeria. And he turned round – they turned round and said 'the camera in the shop saw you take something and put it down your coat' he said 'fine search me'. They searched him and he's going 'you're making me look a crook out here' and they didn't find anything on him ... I went up to him afterwards and I said, it was with my cousin, we went up to him and we said 'look don't worry about it, everyone knows what they're like so you don't look bad', I said. Just I said 'if you want to report it we'll come with you and we'll –' and he said 'no forget it'. And I've seen him around and it like happened to him a few times and he doesn't want to say anything.

It's because if he does say something, it's going to become worser on him if they catch him down Wood Green again because if they catch him this time, they'll either beat him or they'll do something else.

Such positive empathy was quite rare on the street. The only person to volunteer a similar sentiment was an Irish woman who had fostered two West Indian children:

and we were in Budgen's in the town *[in the North of England]* and I'm not kidding you this big bloke walked up in front of me and he said 'you fucking Londoners what are you bringing that up here for' pointing at Joyce, and that really saddened me, put a damper, whereas around here's OK.

The West Indian population was probably the most prominent minority group within the street. But as they did not often impinge upon the stereotype of the 'Black' that arose from the discourse on the dangers of shopping in Wood Green, since (as implied by the concept of cognitive dissonance), no actual Black people are categorised in relation to the concept of the Black found in many peoples' racist discourse. Indeed although there were various incidents and 'troubles' that occurred either as events or as long-term disruptive elements within the street, hardly any of these included young West Indian males. Yet this was precisely the category that was identified as a perceived threat at Wood Green. In short the identification of Wood Green as a Black area is autonomous as a discourse from any actual Black people on the street.

This issue of dissonance becomes still more complex when consideration is given to the Black people who do live on the street. An example is found in Juliette, living on the estate, who was born in Jamaica, was in a long-term relationship with a West African and at the beginning of fieldwork also had an au pair staying with her from Trinidad (although the latter left during the course of the year). She felt and expressed a positive identification with many elements of her identity as Black. This spanned both a specific cultural trajectory from her own Jamaican past expressed in cultural domains such as food preferences or a previous attachment to the Seventh Day Adventist church, but it also spread out to a more generic sense of being Black in Britain evident in her personal relationships and the pluralism of ethnic food in the house.

Her sense of ethnicity occupied a specific component in food shopping, where she supplements the usual supermarkets, with visits to specific outlets for 'ethnic food':

> I usually cook. There is a variety of cooking that goes on in my home. You have the Ghanaian cooking which is palm oil, and then you have my food which is rice and peas and chicken and so on, and then you'll have Dinah Rose's cooking and she does her concoctions, stewed chicken and all the, so there's a variety of food cooking in this house.
> *[How often do you go to get Caribbean stuff?]*
> I'd say weekly, I down there and spend a good £15 on ackee and all that kind of provision.

During a shopping trip to Wood Green we went from a visit to Iceland to an Asian grocer who also stocked African and Caribbean foods. Almost upon entering the shop Juliette completely changed her accent and mannerisms. At no other time in our acquaintance did she let go of a largely Anglicised manner and voice. In the shop, however, she suddenly took on the appearance of a Caribbean market banter. She harangued and hassled the shopkeeper with a mixture of wit and invective that implied the worst possible motives for everything he did but always within an over-the-top and jokey style that made sure that the result would be laughter as much as insult. She shouts: 'Eh what is the price on Carnation evaporated milk? That's too much, I should have got it in Sainsbury's. How come nothing's cold, eh?' She buys a tin of ackee, saltfish, hot Ghanaian pepper sauce, cornmeal, gungo peas – and also the Carnation.

While Juliette is comfortable with her own sense of ethnicity which takes on a highly multicultural sense of Black, that does not lessen the specificity of each of its component regional parts, she has more difficulty with the generic sense of Black that derives from racism but also in part from the younger generation of 'Black' people in the area. She also notes how her own children suffer from such stereotypes:

> Yes there are Black people around here but you know sometimes when I'm studying and that the continual music really gets on my nerves.
> *[Compared to Wood Green?]*
> They're rough and ready in Wood Green, very rough down to earth, well I wouldn't say down to earth but they're rough some of them are rough in Wood Green.
> *[Where do you feel more at ease?]*

Ibis Pond. You don't have to dress up. I could go to Ibis Pond like this, I
mean probably I could go there naked and nobody would pay any attention to me but down that side it's more fashion, fashion is a big thing if
you're not dressing in a particular way it's 'Oh look at how she's dressed',
not that it bothers me, I don't really care, I don't feel so much pressure
in Ibis Pond . . . I feel more free in compared to Wood Green.

Yes I do know people in Wood Green, people who I went to school with
and I'll see them and they'll say, hi – how are you?
[They're into style?]
And they're into style and it's not even the yard guys because I find the
guys are OK, easygoing, but the girls, the girls are very, very aggressive, you know you dare not look on the girls and say even to yourself
'Oh she looks nice' because you'll get a heavy cut eye business you
know so no Wood Green is not really for me, like I said occasionally I
go to Wood Green.

Juliette cannot work with the same cognitive dissonance as non-Black informants; instead she uses a series of registers and levels by which the complex
heterogeneity and reality of people's identities are brought in to create a set
of frames she can use to negotiate her own identity in relation to others, which
include but are not dominated by a negative association between Black and
Wood Green. A parallel situation exists with regard to South Asians and the
associations of the local corner shops. This provides the opposite situation to
West Africans who are relatively free to determine their own retentions or shifts
in identity. In the case of West Indians or South Asians, however, the starting
point is not of their making, but exists in the already present racist discourse
of blackness. Inasmuch as this is partly objectified through the
fear of shopping in Wood Green it is impossible for them to ignore the presence
of this discourse when creating their own relationship with the area, whether
this is to appropriate it as positive or share it but frame it as applying to 'other'
Black people than themselves. In the ethnography as well as in the focus groups
the presence of racism is pervasive and explicit.

The discourse of Englishness

To focus upon Englishness within a chapter comparing ethnicity might be
considered worthy in and of itself in as much as it challenges the simple
dominance of an unmarked majority group that defines all others as marked.
The ethos of this section is, however, based on increasing our understanding
of ordinary people's experiences. In general it must be said that most of the
population do not see themselves as marked or as ethnic, but define themselves mainly by their sense of excluding others as not belonging to them. For
the English the language of identity is based almost entirely on internal categories such as class and gender.

There are, however, some signs that this may be changing, and perhaps more importantly at this stage that a discourse of Englishness is arising amongst these 'others' to which sooner or later the English may have to respond. This concept of Englishness derives from the discourse of what we will call the 'not simply English', by which we mean that although they may have British citizenship and birth, they have a plural identity with a sense of belonging which includes recent ancestry originating outside England. In a sense then the discourse of Englishness has some parallels with that of 'Black' since it arises out of groups other than those designated by the category.

Taking the ethnography as a whole, around a third of informants were not simply English. The term 'informant' refers to the main person within each household who contributed to the ethnography. When partners and others are taken into account then there are many other such groups represented, since partners included several additional Jews, West Indians, Eastern Europeans and others. The general picture is echoed in the local primary school. Languages spoken within this fairly small school included Arabic, Ashanti, Bengali, French, Greek, Gujerati, Ibo, Italian, Kurdish, Luganda, Singhalese, Spanish, Tamil, Turkish, and Urdu, not to mention categories such as 'other African' and 'other Far Eastern'. As within the street the overall picture is one of a very large percentage of not simply English, but with no clear dominant Other. This is the scenario that gives rise to the contemporary characterisation of what we will describe as Englishness.

During the ethnography, Miller's sense of Englishness arose directly from the field experience inasmuch as he had never previously considered the category from the perspective discovered here. Although he is Jewish and positively identifies himself as such, he had not seen this as detracting significantly from his sense of Englishness, which is something he finds himself identifying with naturally in those contexts such as holidays abroad, where a self-consciousness of being English becomes more self-evident. Although he expected to be recognised by other Jews as Jewish on occasion he was genuinely surprised by an experience that occurred several times during the ethnography, which was the direct recognition, based entirely on his appearance, that he was indeed Jewish. Typically he would be talking with an individual such as an Italian, Cypriot or West Indian and at some point in the conversation they would take a long look at him and then state directly that he was Jewish, wasn't he? When he confirmed this identity the point of the question became evident, as they immediately felt free to discuss the nature of the English as a group and ask his opinion on this or that supposed attribute of the English.

His impression is that this is a key development that reflects a locality where there are a very large number of people who are not simply English but whose own group is small and far flung. When someone who is Chinese is chatting to someone who is Argentinian their most obvious point of identity in this context is through their mutual experience of being non-English. The one thing they have in common, apart from the weather and soap opera, is the desire to develop an opinion about the English. By discussing what can be seen as the 'problem' of the English they thereby gain or strengthen their own sense of unity. So while the English discuss the Black and Asian as a problem category, the rest of the population start to develop a counter-category of English as a problem that unites them as a heterogenous group.

As with any stereotype that arises from such a discourse of otherness there are a number of common attributes that soon develop which come to establish for those in conversation that there is indeed such a category and they know of what it consists. In general the discussions focused on what might be termed the 'coldness' or 'emptiness' of the English. It was not that the English were characterised particularly by being exclusive, that is remaining friendly amongst themselves but refusing to allow entry to these others. Rather the English problem was that they were not seen as capable of friendliness even amongst themselves. This was seen as particularly problematic in that people felt that this was not apparently something the English desired for themselves but was rather something that they simply could not overcome. The result of such coldness would be loneliness, depression and aggression to others.

In this regard the main group that objectified the concept of Englishness differed according to the private and public sphere. The discussion about English coldness within the street tended to be based on the working-class English living on the estates. It was they who seemed incapable of becoming friends even with their neighbours, and who then left a whole age-set of pensioners in almost total isolation within the council estates. It was noticeable that one particular council block constantly reiterated its own sense of collective warmth and friendliness by contrasting itself with all the other blocks. Many of the families within that block accounted for this characteristic entirely on the basis that there were virtually no English families on the block, a feature which some saw as a result of deliberate council policy. Clearly then the image of the English working class given in major soap operas was seen as relevant only to the context of the media – not the home. The reality of the English working-class household was felt to be a very long way from *Coronation Street*. The phrase 'this is not Albert Square/this is not Coronation Street' was common, and was intended to contrast the image of Englishness fostered in television soap opera with the inability of the families around them to develop any such larger social contacts.

With respect to the encounter during shopping however, this changes almost entirely to a discourse about the coldness of the middle class. Although this discourse of Englishness varies, in general it is felt that English working-class people may not invite you into their homes but they will chat and smile within a shopping queue or bus stop on the way. By contrast, the English middle class are understood to invite people back to their homes but to be almost incapable of making eye contact or even 'seeing you' when they are out on the street. Within the context of shopping they are viewed as simply rude, competitive and aggressive. This clearly overlaps with similar statements discussed in the previous chapters on class but in particular the chapter on nature.

Although the shopping sites themselves may not be central to this identity formation, in some ways shopping at the two sites may be considered one of the clearest examples of the cause for this emergent phenomena. The experience of shopping at either site is one of considerable heterogeneity of population. Both cover areas with a multitude of different groups and the simple equations of physiognomy such as White and black hide what is a much larger pluralism of regions and languages – a general cacophony of gesture and difference that surrounds people while out shopping. It is likely that as this experience settles into normality it will create new senses of identity. The Cypriot youth manage to construct themselves as a specific group within a generic type – multicultural youth. For most minority people, however, they do not have sufficient size and scope to form such a group and the general sense of plurality is too vague to form the basis of some new identity called British pluralism or multi-ethnicity.

Instead we find that the experience of this heterogeneity may be leading to the formation of a much simpler single identity based around the concept of non-Englishness, in which a unity is constructed around all those who feel free to discuss and debate the problem of the English. For those having this debate it creates through alterity a sense of cohesion which may be felt to be more solid as negation than the potential positive of pluralism. Few people would have the self-confidence to avow the kind of avant-garde lack of locality that is espoused by writers such as Bhabha (1994). Englishness, by contrast, provides a concrete expression of otherness to which they can easily relate, whether it is the idealised English established through chat about characters in a soap opera, or the derogatory stereotypes of rudeness and coldness that are used to characterise the English closer to home.

The importance of the shopping centres may in part be that they are unequivocally major structures of public space. They clearly have the potential to be places where people meet and mix. In many cases it is likely that the people engaged in the formation of this stereotype come from regions where such public spaces are indeed used as public spheres. Many of those from these smaller minority groups are relatively recent immigrants with their own memories of the use of public space in their places of origin. In an ethnographic study which included shopping malls in Trinidad it became evident that shopping in such malls was a largely social activity both in terms of who went but also in the degree of social interaction and banter that takes place within the malls (Miller 1997: 293–300). To the extent to which this is a common background, then, the shopping centres become a key emblem of the 'problem' of the English in that these are sites where the lack of social interaction is particularly conspicuous, when compared to the use of such public spaces in other countries.

The other possible direction that might follow from these observations about the development of a general discourse about Englishness is that the concept

181

may start to impinge upon those who are thereby stereotyped in the same way that the concept of Black becomes something actual Black people must choose to appropriate or defy but cannot ignore. To date the two formations are unequal since Black is foisted on the dominated by the dominant, while English is still a tentative formation by the weak against the majority, who are mainly quite unaware of its existence. But at a time of almost unremitting political and economic decline over several decades, when previous formations of positive identity centred on the concept of being British are becoming ever more fragile, it is a reasonable prediction that this formation of Englishness may well become central in the future to any new attempt to fill the empty space of identity that has emerged amongst the dominant population. At present, however, the Englishness of shopping centres is for the 'simply English' mainly a perception that exists in contrast to anecdotes about shopping abroad while on holiday.

Conclusion

The overall result of our research into different groups' relationships to the two centres is that these are not a particularly important source of identity for those who live on either end of the spectrum of ethnicity that has been outlined here. In the case of the West Africans one has a given sense of culture so that there is no need to use this additional locality to construct the notion of identity. This would also hold for many other groups in a similar position. At the other end of the spectrum is the concept of Black which is derived from a racist discourse to which the group so defined may respond either by trying to appropriate the category and turn it into positive identification as in Black Power, or by avoiding it and excluding it from a variety of regional identities that are self-designated such as Guyanese or Punjabi (Modood 1988). The shopping centre as locality in this case becomes a site of Otherness, and it becomes associated with Black as threat of crime felt to lurk in the environs (especially the associated car parks). There are parallels with the concept of the English as a problem, one facet of which is their behaviour as encountered while out shopping, but this has not yet led to a strong or explicit sense of the Englishness of the shopping centres, though this may occur in the future.

The real importance of these sites as an objectification of the specific sense of identity emerges for those groups that lie midway between these two positions. The Jewish group's fear is precisely that they have lost or are in danger of losing the taken-for-granted sense of identity that comes from given origins and which therefore affirms them without as it were any work involved in this construction. On the other hand they wish to retain a sense of what it is to be Jewish since the image is not simply the creation of racism but a positive concept of commonality based on religious identification and practices which demand separation, such as endogamy (Anthias and Yuval-Davies 1992). In this case, then, there is an intense striving built up in little stereotypes which appropriates Brent Cross from its general non-ethnicised context and from that extracts a specifically Jewish Brent Cross which in turn can be used to affirm a specific sense of at least one variety of being Jewish.

A version of the same logic applies to the Cypriot youths' use of Wood Green. Here it is not a general sense of Cypriot which is at stake, since the older generation are sufficiently close to their point of origin to take their identification from a sense of such origins rather than within the British context. The younger generation, by contrast, is no longer in a position to do this, and furthermore is wishing to assert an identity that distances itself from its own parents. In this context Wood Green Shopping City is the focus of the Byriters who affirm a specifically British concept of themselves as an ethnic minority with affinities to other ethnic minority youths, but also a specific grouping based around the site itself.

The conclusion which must follow such observations is that one cannot hope to create any simple theory of the relationship between ethnicity and locality. Rather 'ethnicity' exists as a wide range of identifications which may be generated by those within a community or projected onto that community by others. Earlier studies of ethnicity tended to concur with an essentialist and normative position that assumed the subjects of study. Thus all people of Cypriot origin are 'Cypriots' and we can generalise them as a group regardless of their own attitude to this aspect of identity. Modern studies attacked these assumptions and demonstrated the degree to which individuals may not conform to such assumptions and focused rather on the 'Apache Indian' potentiality of pluralism. The emphasis is on a state of creative appropriation of the possible interweaving or rejection of given ethnicities. From this emerges the current tendency to evoke postmodern theory behind an assertion that identity has become a general and rather vague set of plural possibilities which individuals are free to pick up on and appropriate in various and creative ways.

The evidence of this study, however, is that ethnic labels and categories are used here because they are clearly avowed both by those within them and through racist projection onto groups by others. There are considerable limits on creativity and appropriation. What the focus groups and ethnography suggest is that there are indeed groups of people who have very firm relations of identity as young Cypriots or Jewish women which are highly normative and bounded, and increasingly objectified through the medium of the shopping centres themselves. But the degree of identification appears almost unrelated to any particular quality of that group. It does not depend upon any claim to its 'cultural' attributes, and is better related to its historical trajectory as a minority. The tradition behind this account is then closer to Barth (1969) (though without his stress on individual choice) in seeing ethnicity as a contextual construction that transcends any list of attributes. As such this form of identification does not appear particularly modern, let alone part of a postmodern 'tribal' and transient form of sociality (see Shields 1992b: 99–113; Maffesoli 1996).

At the same time, there are others who have a strong sense of ethnicity but for whom the locality is quite irrelevant, or for whom the locality is important but ethnicity is not. There may be still others who cannot engage in creative appropriation because the relation of site to identity is constructed by others and projected onto them. Banks (1996: 182–90) has recently reviewed the literature on ethnicity and urged caution in presuming this term outside of the heuristics of academic analysis. The groups described in this chapter may not consider themselves ethnic, but they certainly do consider themselves Cypriot, Jewish, West African, etc. Ethnicity here is the generality that allows these identifications to be compared and contrasted. But even these separate identities cannot be presumed. Our starting point is the observation of their use, where projected, where repudiated and where appropriated. Our concluding point is an affirmation of the spatial dimension represented here specifically by shopping centres not merely as a locus of ethnicity but as an important medium through which some people's identities are formed and practised.

9

CONCLUSION

A material culture approach

Throughout this book we have attempted to ground our understanding of the relationship between shopping, place and identity through detailed empirical research in two North London shopping centres. As a result, our findings contrast with some of the more ungrounded speculations about contemporary consumption which have tended to see 'identity' as inherently plural and free-floating and 'consumption' as a hedonistic pursuit of a virtually limitless range of lifestyle choices. Our approach has been rooted in the material culture of specific places: the shopping centres themselves and the neighbourhoods and communities around Brent Cross and Wood Green. Through shopping in these particular places, consumers are involved in a creative reworking of gender, ethnicity, class and place. Whether shopping on their own or with others, they are making significant social investments in a relatively narrow set of family and domestic relationships as well as making economic choices about the utility of particular goods. In this final chapter we draw out some of the more general conclusions of our research in terms of the mutual constitution of place and identity, the advantages of using multiple methods, and the significance of our findings with respect to future policy and future research.

The tenor of our analysis has been to refuse the autonomy of either place or subject, and rather to use this project to examine the articulation between identity and place and thereby transcend any simple duality of subject and object. Although topics such as ethnicity and gender are central to our analysis, we do not take these as pre-given social parameters of identity that are then subject to some process of symbolic representation by place. The previous chapters demonstrate instead that the very materiality of place itself becomes a medium for the objectification of ethnic, gender and other forms of identity under certain conditions. That is to say that place is itself partly responsible for the form that they take.

To become a Byriter is clearly of some significance to the Cypriot youths who live in the Wood Green area. It provides one of the main sites of identity for this group and thereby a potential statement of their cultural development.

The category of Cypriot youth may include some for whom this has become the linchpin of their assertion of a new identity which takes its meaning from the space in which it is manifested. It is equally possible for other sections of Cypriot youth to reject this option for a variety of reasons in favour of other possibilities. Even for them, however, the shopping centre stands as the locality for an identity in opposition to which they are defining themselves. As a result the experience of shopping, especially for them, but to a lesser extent for all shoppers at Wood Green, is transformed by the evidence that Shopping City has become a site for the collective assertion of a particular identity by some of this group.

In such a case identity and place become mutually constituted. Cypriot identity is not being reflected here, but is given a new meaning quite distinct from any previous form and one that takes its presence from the sense of place. This is not to say that the relationship cannot be generalised and theorised. The point of the previous chapter was to suggest that if the range of potential ethnic identities is considered, some may have no relation to such a place, while others become increasingly associated with it. We suggest that generalisations may be made about the particular stage in the development of the relationship between a group and its identity within British society which makes it more or less likely for this kind of process to occur. The result is not, of course, a necessary relationship.

Other groups may not use the particular possibility of shopping spaces in the way that was observed for Cypriot youths in Wood Green and Jewish women in Brent Cross, even if they are in other respects in parallel situations, but the generality at least helps us to understand why this specific medium of objectification may have become related to some groups rather than others.

Although it is evident that such a relationship may be refused as well as accepted by an individual, the very materiality and scale of place constrains the possibilities for individuals and creates normativity – what we think of as typical responses – as an outcome of its presence. This is evident from the chapter on class (chapter 7). The analysis of class suggests that the two sites we studied may become highly structured objectifications of the meaning and experience of class in the contemporary world. That was demonstrated through our ability to locate a systematic opposition between the values expressed in each case such that either is partly defined by its opposition to the other. So the rational order of John Lewis is held against an opposed image of order in the cheap-jack based upon taking advantage of fortuitous supply. It was not simply that the sites themselves represented class. On the contrary, we demonstrated that in most respects the shopping areas were very similar. It was rather that the consuming population utilised certain key features of the two centres to break them apart, as it were, and manifest an opposition sufficiently systematic as to appear within analysis as a structural opposition. This means that the sites can represent either class diversity or class similarity depending upon whether or not these particular shops become important to a shopper.

The same argument that is made for place can be made for identity. We suggested that amongst the informants there are those whose identity in terms of class is clear, unequivocal and monolithic. In such cases there is nothing plural or multivocal about them. They identify with their respective site of John Lewis or the cheapjack and deride, or are intimidated by, that with which they do not identify. There is nothing here to confirm recent theoretical speculation that all modern identity must be multivocal. At the same time there are many people who are comfortable with both experiences of class because they empathise with both as dual aspects and potentials of themselves. They do not experience or understand this as a contradiction, and even though the outside observer notes the remarkable changes in their style and manner as they move between such contexts, this 'code-switching' is natural and unconscious to them.

This does not mean, however, that place is merely a site for the performance of pre-given genres of identity. On the contrary, our argument is that the material culture that is a shopping site becomes itself a form through which the nature of such identity is discovered and refined. The experience of the cheapjack reminds people of what a particular kind of image of working class can be, and allows them to feel at home in its presence and learn its subtleties and ironic humour. Shopping at John Lewis in turn acts as a kind of material habitus in that, as Bourdieu (1968) noted of such material taxonomies, it is not any one feature, such as design, rationalism, value, compromise, etc. that creates this disposition but the structural homology between a whole set of compatible values expressed in the taxonomic orders which together come to be recognised as the particular John Lewis experience. It is this habitus, where one set of values tends to reinforce all the others, which becomes in turn the quintessence of a particular range of values that we understand as class.

We do not then encounter some free plurality of agents 'choosing' their class. We find instead a highly normative structure of oppositions based on a complex interlocking package of variables which allow individuals only one of three positions, to identify themselves with one site, with the other, or with both. But the constellation of values represented as the experience of class is well established in the cultural conventions that are expressed in the shopping forms themselves, and do not therefore exist merely within people's consciousness. Class here is a practice, that is a processual relationship between persons and place. Its objectification in shopping is not unrelated to class as an expression of occupation, or of aspiration, but shopping provides a particular structure of difference that is not going to be quite the same as any other expression of class and therefore adds its own contribution to what we understand and experience class to be. Material culture is not something that exists just in the materiality of place or objects. Nor is it just an internalised fantasy of these places or objects. It is called material culture because it exists in the practices which involve both these things.

Interest groups and agents

The places of concern here are not some autonomous trajectories in history. They should not be treated as 'discourses' or the mechanical results of a historical process reflecting larger 'fashions' and structures. They are certainly of their time, and provide evidence of the values and structures of that time. But this is because of the active work of specific groups of people who have particular interests involved in the creation and maintenance of these places. In particular we are concerned with managers, planners and owners. In understanding the articulation between the shopper and the place of shopping, therefore, the relationship between the producer and consumer of space has to be observed and analysed in itself. This is not at all to suggest that the place is merely the manipulated expression of the intentions of management. We do not presume, for example, that the management would perceive John Lewis or the concept of nature in the way we have represented them within this volume. Many of the developments take place in the context of larger relationalities of which they are only partially aware. It is quite unlikely, for example, that a Brent Cross manager is thinking in terms of an opposition to the Wood Green cheapjack since they are focused largely on what for the shopper is only one genre within a much wider range of shopping experiences.

Chapter 2 argued that behind these two sites are quite different histories, with Brent Cross essentially the creature of commercial interests co-opting others such as the local council where there is mutuality of interest but being driven by its central commercial imperative. Wood Green, by contrast, was initially conceived of as part of a strategy of locality regeneration, and has always been in part a commercial venture that had to be also an instrument of local government. Still today its future is partly the concern of people in posts such as 'town centre manager', and Haringey council continues to see the site as an instrument of suburban renewal. These differences in the histories and intentions behind the sites are important in understanding their contemporary role and their relationship to the community of shoppers.

Similarly, the effect of these places in the contemporary objectification of class discussed in chapter 7 becomes clearer when set against the larger history of the use of the department store to construct the values embodied in middle-class identity (Lancaster 1995). We thereby appreciate the long time-scale in which the department store has developed its facility to express this particular range of values, and more recently to make these accessible to a shopping community that is as much working class as middle class. The same larger history is needed to incorporate the other perspective which is that of retail employees. Although we did not study shop floor workers directly the relationship between workers and shoppers became a core element to the discussion of chapter 6, because retail had been a major employer of those whom we encountered as shoppers. This historical relationship in turn proved to be a major distinction dividing attitudes to the shopping centres along class lines according to whether shoppers tended to consider shop workers as part of their own community or as a 'natural' part of the shopping environment.

Clearly, we need to attempt an understanding of those ideological premises of management, workers and shoppers that may have consequences of which none of these groups alone is fully conscious. It may or may not be the case that management are aware that the concept of family shopping that they use to sell the shopping centre to the shopper is largely an ideological construction that ignores the fact that most shoppers do not like to shop with their families but can stand for an assurance of security against various fears, including racialised prejudices which the shoppers may be unwilling to have explicated even by themselves. The effective collusion between commerce and consumers to represent aspects of shopping such as 'nature' in a manner that bears little relation to any larger understanding of the term but is sufficient to assuage the fear of being seen as rampant materialists does not depend upon either commerce or the consumer having to come to terms with the contradictions in their position. Indeed, it is the ability of place as material culture to resolve contradictions in ideology without these necessarily coming into consciousness that makes them so effective as modern-day 'myths' in the sense of Barthes (1973).

Our work as academics has been to try to excavate and reveal these same contradictions which the actors in the world of shopping have no particular reason to express. This does not make us either ignorant of, or less respectful of, the agency involved, whether in commercial decisions or shoppers' choices, but we simply do not regard either agency or choice as some simple manifestation of a pre-given condition to be called identity. On the contrary, our main concern has been with identity as the consequence rather than the precondition of shopping, in the same way that we want to understand place as the effect and not the pre-given premise of the shopping encounter. A material culture approach allows us to avoid such simple affirmations of either place or identity.

Multiple methods

It is no coincidence that a project based upon a multiplicity of methodologies has ended up with a focus on contradiction in representation and identity. The claim we would make is that the use of a range of quantitative surveys, qualitative focus groups and ethnographic observations is fully justified by results which show that these methods can be both complementary and contradictory. These are equally useful conclusions. On the one hand, the larger base of the quantitative survey complements the ethnography and focus groups and assists any generalisations being made from the research. To make general statements about which category of people tends to shop where requires this larger number of informants. The survey also provided the basis for the selection of the areas for the more focused research. The ethnography was selected on the basis of a map, plotting the postcodes of those who had answered the questionnaire. On the other hand, where the evidence which emerges from the these different approaches comes into conflict, then, as made clear in the chapter on policy (chapter 4), we have seen this as an additional and important result of the research that transcends that of the individual methods.

Different methods lend themselves to the creation of different perspectives which are equally interesting when they provide supportive results in which one thereby has more confidence, where they provide complementary results, and when they show a clear contradiction in the conclusions that one would have drawn from them. Much of the overall description of the shoppers that emerged from the quantitative questionnaire conformed to that which emerged from the focus group and the ethnography. The generally benign view taken of both centres was evident in all three forms of enquiry, as were the overall characterisation of the two localities. Complementarity is best shown where there were forms of understanding which simply could not arise from one or other method. The previous chapter on ethnicity is taken almost entirely from the focus group information. The reason for this is that although the ethnography contained a large number of people from a wide diversity of backgrounds, there was no single group sufficiently represented that one felt able to generalise about the relationship of such a group to the shopping centre, with the exception of the overall attitude to the English, which is why this is the only evidence from the ethnography that is employed in chapter 8. In short, we would not have been able to write a chapter on this topic on the basis of either the ethnography or the survey. It was only the focus groups which gave evidence from groups of people who had already affirmed their association with particular identities and whose discussions as a group allow one to see what attitudes are being held in common in terms of that particular identification.

In addition to confirmation and complementarity is the evidence for contradiction between our various methods. Our discussion of the concept of family shopping emerged from a discrepancy between discussions with both management and shoppers, and the evidence from the survey that showed that almost no one actually wanted to shop with their families. The clearest evidence for contradiction is that given in the chapter on policy (chapter 4) where it was shown that the evidence taken from elderly people in focus groups is directly contradicted by the evidence of elderly people on their own or when actually observed in shopping. One of the points made in that chapter is that we do not thereby conclude that one kind of evidence is false or less reliable than the other. Instead we argue that the contradiction is itself valid, that the same people feel differently about a place or activity when in the context of a group that stresses their identity as elderly people as opposed to a context of their individual relationship to these activities. The use of multiple methods is not therefore some kind of check on the 'truth' of informants, as though we were engaged in a version of detective work. Rather it helps us develop a more nuanced understanding of the way the same people incorporate a variety of perspectives and attitudes which are brought out by different contexts. This relates back to the earlier point about material culture, in that it affirms that it is often the context of elicitation, the place itself, that helps determine the identity of the subjects being studied.

Policy and politics

This stance also explains our response to the debates over the politics and policies of retail. We do not wish to detract from the task of politicising retail, that is, the attempt to insist that shopping has larger consequences than that to which the commercial interests involved would necessarily wish to draw attention. Thus for example when Raven et al. (1995) insist that the centralisation of supermarkets has implicated a change in road haulage which may have detrimental consequences for us all that have been factored out in considering the 'efficiency' of such retail developments, this seems an entirely reasonable accusation and broadening of any debate as to the proper determinant of future retail policy.

What we have criticised, however, is the tendency to project onto populations an ideological stance to fit within such arguments. Those who favour market deregulation tend to construct a pure consumer whose only interest is in a greater range of choice and lower prices. The shoppers studied here, by contrast, tend to have a much wider range of concerns, and perhaps surprisingly it may be that the most important determinants of shopping are social relations and morality (see Miller forthcoming). On the other side we find that those who oppose new developments tend to homogenise groups such as the impoverished, the elderly or the disabled and reduce them to a kind of functionalism of basic need, in which they ought to be against this or that change. In some ways this apparent radicalism invents its own conservatism, of nostalgia based on a romanticism of the previous forms of retail which denies the evidence that this is not shared by those for whose interests they claim to stand. The elderly may be nostalgic, but equally they may turn out to be the most radical of modernisers precisely because they remember the problems and suffering associated with what others now romanticise as past practice.

Our research revealed the complex articulation between factors which help to determine benefits at particular times. We might conclude from a simple view of shopping interests that out-of-town large-scale retail is preferable as long as there are good transport facilities for the poor and elderly. But if this means that the emergency provision provided by the corner shop is eliminated and the retail areas of inner cities become destitute and unsafe then attitudes to shopping are clearly too narrow a focus to be the sole arbiter of shopping policy. Future policy might better work towards transcending the current divisions between these two retail forms and work instead to obtain the benefits of scale that produce a greater range of goods at lower prices directly to consumers who have difficulties in reaching large out-of-town facilities. Our particular sites as suburban shopping centres also help break up too simple a distinction between inner-city and out-of-town shopping developments.

The two shopping sites we studied are not typical of any particular phenomenon. They were selected as contrastive. We have noted the different political context and relationship to locality that makes them very different sites. Although they are both suburban shopping centres, Brent Cross carries the unique label of Britain's first mall and is currently engaged in expansion and continued success. It may well be that the singularity and clarity of identity, occupying what had been a missing position as the North London local alternative to West End shopping, has itself played no small part in its success.

Wood Green, by contrast, has had a much more ambivalent relationship to both place and identity. It never occupied the city centre position of most similar shopping centres, and even with respect to the High Street around it, its position is seen by the public as unclear. While Brent Cross manages to combine the modern trajectory towards out-of-town car-based experience with the position to rival city centre shopping, Wood Green has failed to relate clearly to either of these alternatives. This may well be the reason that it has been rather less successful as a commercial proposition. This is also related to contingent factors such as the relative decline in the status of the area surrounding Wood Green itself. This does not suggest that the successes and failures of such centres can be regarded as entirely a matter of good or bad management or investment. As we have tried to demonstrate, many of the most important changes have to do with larger changes in popular ideologies. These may have to do with factors such as the concept of nature, the manner in which we develop the term 'family' as metaphor, the conservatism of racism and the general appropriation of different public spaces to objectify tensions in many other aspects of identity such as gender and class. These may not be the major factors which come into discussion in explicit discussion of politics and policy. Equally, we do not expect that commercial interests have a particularly clear sense of such developments, and even as academics dedicated to this precise topic we recognise that what we encounter is a complex articulation between a wide array of moral and social values, some with long histories, but no more predictable as to their future trajectory for all that.

Future directions

What has emerged through our research is the problem of an overarching ideology that projects a unity of goods and people. The elderly who are fighting to see themselves as young and modern do not wish to be associated with 'olde worlde' shops and goods that are the contexts in which others would wish to place them, just as children gain their autonomy by rejecting images of childhood that parents and commerce collude to create as the proper context for modern childhood. Our research suggests particular foci for future investigation. The impoverished elderly living on council estates emerged as the most clearly disadvantaged by traditional retailing. This suggests that their welfare should be a primary concern for those involved in the development of new retail technologies such as television-based selection of goods which are likely soon to be making a significant presence. These new technologies might exacerbate or alleviate the problems revealed by our research depending upon whether the interests of the elderly are taken into account in their early development.

While we can discern particular needs uncovered by the research, we do not wish to subscribe to a simple populism that implies that techniques such as ethnography allow us to stand more directly for the consumers whose authority is absolute. One of the main themes of chapter 5 is that current changes are in part based on a façade of family values that masks a highly racialised discourse of locality that is held by shoppers. This authenticity of racialised perception makes it no less detestable. We do not then naively stand by the voice of the 'true' consumer whose interests we now claim to better represent. On the contrary, the whole tenor of our investigation is to suggest that there is no such thing as the consumer, either the individual consumer of economic theory or the denied consumer of radical critique. If the identity of the person is in part lodged in the very places themselves, such as shopping centres, then both the autonomy of subject (identity) and object (shopping centre) is refused, and what academic encounter reveals is rather the contradictions which are thrown up by the articulation between person and place within which both person and place are forged. The resulting picture may be far more complex, but the role of the kind of independent academic research that we have undertaken is precisely to reveal these complexities rather than allow them to be brushed underneath a carpet of pragmatism or ideology. Indeed, the most thorough of the reviews of the research literature (BDP Planning and OXIRM 1992) can be found on close reading to have uncovered a similar level of ambivalence and contradiction in shopping centres, though with less attempt to theorise these than has been made in the current volume. What we would assert, however, is that notwithstanding these complexities, it is possible at a given moment of time to investigate the manner in which particular public spaces such as shopping centres not merely reflect but play an active role in the objectification and thereby in the transformation of the values and identities of shoppers.

NOTES

1 CONSUMPTION AND SHOPPING

1 Even in countries such as Japan which have notorious concentrations of producer power, retailers are now making their presence felt.

2 There are of course other accounts that we might have pointed to. For example, there is an account which stresses the rise of ethical shopping. Of course, ethical consumerism has a long history (for example, see Haynes' 1983 piece on the fight for the Plumage Act of 1921) but it has undoubtedly become more important of late. Another 'cyber-account' would have stressed the rise of the internet and its potential as a means of obtaining an almost infinitely mobile choice. For Mitchell (1995), for example, shopping is about to be struck amidships by information technology and the result will be radical reordering of its conduct: Electronic Fund Transfer at Point of Sale (EFTPOS) systems are just the start; soon there will be virtual reality stores. Product suppliers, sales persons and customers will no longer have to come together to make a sale, and the nature of shopping will therefore radically change.

With such soft shops, specialised retail districts and the departments that make up department stores will simply become directory categories – logical groupings presented as menu items, or virtual storefronts in the interface of on-line services. Retail location will become a matter of being in the right directories.

As with the old Yellow Pages, customers will let their fingers (or rather, their cursors) do the walking. The stock will be bigger and the selection larger than in the mightiest big-box off-ramp superstore. The things that will remain in physical form are warehouses (which may become smaller as computerised inventory control strategies become more sophisticated) and delivery vehicles.

From K-mart to Cybermart! *Sic transit* retail space?

One more account might have stressed the difficulties of shopping centres for relatively immobile sections of the population. For Robillard (1996), for example, shopping is about anger born of the frustration inherent in being unable, as someone who suffers from severe muscular dystrophy is thereby confined to a wheelchair and is only able to speak through a 'translator', to orchestrate the social encounter. Shopping then consists of botched attempts at encounter which negate the social engagements Robillard wants to have:

> I was at the end of a shopping trip to the local mall when an old friend, a former sailing buddy, approached me from behind, placed his hand on my back, saying 'I have not seen you in a long time!' ... I wanted to speak. But he moved so quickly to my left side, while I was looking to the right to find a translator, that the exchange and facial contact needed to

initiate a conversation was barely apparent ... After his brief greeting, my friend bounded up the stairs and was gone before I could make the appropriate moves to start a conversation. I was seething with anger.

(Robillard 1996: 20)

Why, asked Robillard, do people find the sight of him in the mall so difficult to cope with? Because, he replies,

I am seen as a person paralysed, immobile to the interactional surroundings. The mere sight of me, as well as other disabled people I have observed in malls, often generates head-shaking, scowls, even audible cursing. It is reputed this decision stems from fear, that 'There go I but for the grace of God' syndrome. It is said that witnessing a disabled person is equivalent to seeing one's own mortality. I think these reasons have little to do with the decision. I surmise that in the perception of others one sees the full range of bodily instrumentalities, calling out and institutionalising, moment by moment, one's own body capacities and opportunities, and the sight of the paralysed, the crippled, the lame is a sharp denial of this common-sense reciprocal knowledge.

(Robillard 1996: 22)

3 Thus Freud wrote 'I don't hate America, I regret it!' America stood for passionless conformism and impoverished materialistic ideas.

4 Samuel (1994) notes that 80 per cent of Past Times customers are women.

5 Thus, Finch (1989: 84) writes: 'if there has been any discernible change in a single direction over time, it has been an increase rather than a decrease in the significance of the feelings which people have for their relatives [the particular blend of instrumentality and expressiveness for given individuals is a matter for empirical investigation]' (See also Finch and Mason 1990).

6 A more relational approach to identity may also counter the excessively individualistic implications of some forms of 'identity politics' such as those that Eric Hobsbawm (1996) has recently attacked. Arguing that the current preoccupation with consumption represents a triumph of rhetorical style over political substance, Hobsbawm contrasts *sectional* politics (based on particular identities) with *general* politics (based on the alleged universalism of the Left), arguing against a vision of the Left as 'a coalition of minority groups and interests ... [which] hardly ever mobilizes more than a minority' (1996: 44).

7 Paul du Gay's (1996) analysis of the discursive construction of new work-based subjectivities theorises identity in a similar way, as contingent and relational.

8 In 1995–6, 44 per cent of the revenue of BAA plc, the company that runs most of Britain's major airports, came from retailing.

9 We might add that malls are often anchored by one or two department stores, a nice illustration of how older (though suitably modernised) retail forms co-exist with the new.

10 And this is to ignore the second-hand goods market. By value alone, this market is massive, most especially because of the contribution of second-hand cars, yet, even now, it is remarkably little studied. Perhaps this is because second-hand goods are seen as historically more important than now. If so, this judgement is incorrect.

2 HISTORY AND DEVELOPMENT OF BRENT CROSS AND WOOD GREEN SHOPPING CENTRES

1 In fact, the two malls that we have studied are unusual in the sense that they are not out-of-town. Wood Green is clearly an urban shopping centre, while Brent

Cross, 'in reality, is North London's missing major retail area and has to be viewed as one of London's four major suburban centres, along with Croydon, Kingston and Romford' (Lancaster 1995: 199).

2 Hammerson and Standard Life are closely tied together. Standard Life has been an investor in Hammerson and its developments since the 1950s and in 1989 protected Hammerson from a hostile bid from Rodamco, the Dutch property investment fund.

3 Security is a key issue for two reasons. First, because customers clearly want it. And, second, because there are some real dangers. For example, in 1992 Wood Green Shopping City was the subject of an IRA bomb attack. In 1991, four bombs were discovered at Brent Cross, two of which went off, causing fires at C&A and Fenwick's.

4 There are other examples of this division of labour. For example, there is the rise of the facilities manager, in charge of all the operations of a commercial office block. And there is the rise of the town centre manager. There were only nine of these managers in 1991. Now there are 160. A city centre manager was appointed for Wood Green in 1993, on secondment from Marks & Spencer, as part of a community initiative.

5 Of course, retailing ideas can travel the other way. For example, the expansion of grocery stores such as Sainsbury's in the United States has brought with it practices from Britain (see Wrigley 1997a, 1997b).

6 There is a by-law which states that John Lewis must have an uninterrupted view of Fenwick's (and vice versa) with nothing over 3ft 6in tall in between.

7 For example, Brent Cross has had the problem of larger retail chains coming in and taking over from independents, thereby reducing the amount of space for high-class clothing boutiques. In part, the new extension is meant to solve this problem.

4 SHOPPING POLICY AND SHOPPING PRACTICE

1 Another recent study of safety and shopping (Thomas and Bromley 1996) showed consumers in Cardiff and Swansea city centres to be fearful of streets, car parks and transport termini because of environmental deterioration, anti-social behaviour and problems with the security of their cars.

2 According to the MINTEL report on consumer shopping habits (MINTEL 1994: 27) only 10 per cent of those over 65 claim to shop regularly at out-of-town centres, compared to 17 per cent for the 35–44 age group. The only category where the elderly score higher than any other group is the market. When these data are combined with information on socio-economic group, however, the numbers of the elderly in the higher groups who shop at out-of-town facilities are almost twice that of lower groups (ibid.: 31).

5 FAMILY SHOPPING AND THE FEAR OF OTHERS

1 For the history of the more general incorporation of the crowd into the interior worlds of shopping without threatening the imagination of community, see Slater (1993).

2 See, for example, Rosalind Williams' (1982) discussion of mass consumption in nineteenth-century France, Gail Reekie's (1993) study of the sexual temptations offered by the early department stores in Australia, Judith Williamson's (1986) study of the seductive pleasures of popular culture, and Victoria de Grazia's (1996) study of 'sex from things'.

3 As we shall see, those consumption spaces that are most often associated with men (such as DIY or garden centres) or consumption activities in which men engage as frequently or more so than women (such as the purchase of cars, houses, books or

computers) are rarely even regarded as 'shopping' in the conventional sense. In everyday language, 'shopping' is usually restricted to the purchase of food and clothing, stereotypically regarded as 'women's work'.

4 Recent marketing surveys suggest that women make up 60–65 per cent of consumers at most shopping locations, with a higher proportion of male customers at out-of-town centres (MINTEL 1994: 5). The same report suggests that 'some 15% of adults frequently visit shopping locations primarily for pleasure' and that 'Men enjoy shopping less than women' (ibid.: 6). We would suggest that a more nuanced picture of the gendered nature of shopping is needed, where the alleged pleasures of shopping are more carefully contextualised.

5 Moreover, the ethnographic evidence suggests that many men go shopping under the direct guidance of their female partners who provide lists or instructions for them. This allows men to undertake the actual activity of shopping while distancing themselves conceptually from the idea that they are a 'shopper'.

6 These results cannot be accounted for in terms of the family status of respondents as 18 per cent of respondents at Brent Cross and 20 per cent at Wood Green had dependant children living at home.

7 The 'nightmare' metaphor was used by several other women at both centres: 'going to Waitrose at Brent Cross is a nightmare', 'there's a nightmare of traffic outside', '[the lack of public lavatories] just makes shopping a nightmare' and 'The market [at Wood Green] is a nightmare if you've got a pushchair'.

8 Sharon Zukin (1995: xiv) uses a similar metaphor in her analysis of the gentrification of Bryant park in New York which she describes as 'domestication by cappuccino'. While she does not develop the implications of the metaphor at any length, it has a range of meanings all of which are relevant, in varying degrees, to the processes described here. These include the taming and civilisation of wild or savage environments, the privatisation or interiorisation of space, the enclosing and private ownership of land, and the feminisation of nature, making a home or familial place from what was previously foreign or hostile territory (see Jackson 1997).

9 Declining crime levels were reported at a Wood Green Open Forum addressed by the local police on 4 September 1995 and were subsequently publicised by the Town Centre Manager, Richard Thomas (interview: July 1995).

10 Les Back (1996) describes a similar sense of nostalgia among White residents in London's Docklands whose neighbourhood had been subject to a comparable process of 'racial' change.

11 For an anthropological argument about the meaning of dirt ('matter out of place'), see Douglas (1969).

7 JOHN LEWIS AND THE CHEAPJACK: A STUDY OF CLASS AND IDENTITY

1 In addition to the two shopping centres, reference is also made to a local shopping area called here Ibis Pond, which is if anything more unambiguously associated with being middle class than either of the two shopping centres studied.

BIBLIOGRAPHY

Abelson, E.S. (1989) *When Ladies Go A-Thieving. Middle Class Shoplifters in the Victorian Department Store*, New York: Oxford University Press.

Allan, G. (1996) *Kinship and Friendship in Modern Britain*, Oxford: Oxford University Press.

Anthias, F. and Yuval-Davies, N. (eds) (1992) *Racialized Boundaries: Race, Nation, Gender, Colour and Class and the Anti-Racist Struggle*, London: Routledge.

Anderson, B. (1983) *Imagined Communities*, London: Verso.

Appadurai, A. (ed.) (1986) *The Social Life of Things*, Cambridge: Cambridge University Press.

Augé, M. (1995) *Non-places: Introduction to an Anthropology of Supermodernity* (trans. J. Howe), London: Verso.

Back, L. (1996) *New Ethnicities and Urban Culture*, London: UCL Press.

Banks, M. (1996) *Ethnicity: Anthropological Constructions*, London: Routledge.

Barth, F. (1969) *Ethnic Groups and Boundaries*, Bergen: Universitäts Forlaget.

Barthes, R. (1973) *Mythologies*, London: Paladin.

Baudrillard, J. (1988) 'Consumer society', in M. Poster (ed.) *Jean Baudrillard. Selected Writings*, Cambridge: Polity Press, 26–43.

Bauman, Z. (1993) 'From pilgrim to tourist – or a short history of identity', in S. Hall and P. du Gay (eds) *Questions of Cultural Identity*, London: Sage, 18–36.

BDP (Building Design Partnership) Planning and OXIRM (Oxford Institute for Retail Management) (1992) *The Effects of Major Out-of-Town Retail Developments: A Literature Review for the Department of the Environment*, London: HMSO.

Beck, A. and Willis, A. (1995) *Crime and Insecurity: Managing the Risk to Safe Shopping*, Leicester: Perpetuity Press.

Beck, U. (1992) *Risk Society*, London: Sage.

Beddington, N. (1991) *Shopping Centres: Retail Development, Design and Management*, Oxford: Butterworth Architecture.

Belk, R. (1995) *Collecting in a Consumer Society*, London: Routledge.

Benson, J. (1994) *The Rise of Consumer Society in Britain 1880–1980*, Harlow: Longman.

Berman, M. (1986) 'Take it to the streets: conflict and community in public spaces', *Dissent*: 470–94.

Berman, M. (1988) *All That is Solid Melts into Air: The Experience of Modernity*, London: Penguin (first published 1982 in New York: Simon & Schuster).

Bhabha, H. (1994) *The Location of Culture*, London: Routledge.

Billig, M., Condor, S., Edwards, D., Gane, M., Middleton, D. and Radley, A.R. (1988) *Ideological Dilemmas. A Social Psychology of Everyday Thinking*, London: Sage.

Blaut, J. (1992) 'The theory of cultural racism', *Antipode* 24: 298–9.

Bloch, P.H., Ridgway, N.M. and Nelson, J.E. (1991). 'Leisure and the shopping mall', *Advances in Consumer Research* 18: 445–52.

Bocock, R. (1993) *Consumption*, London: Routledge.

Bourdieu, P. (1968) *Outline of a Theory of Practice*, Cambridge: Cambridge University Press.

Bourdieu, P. (1984) *Distinction: A Social Critique of the Judgement of Taste*, London: Routledge and Kegan Paul.

Bourdieu, P. (1985) 'The social space and the genesis of groups', *Theory and Society* 14: 723–44

Bowlby, R. (1985) *Just Looking: Consumer Culture in Dreiser, Gissing and Zola*, London: Methuen.

Bowlby, R. (1993) *Shopping with Freud*, London: Routledge.

Brewer, J. and Porter, R. (eds) (1993) *Consumption and the World of Goods*, London: Routledge.

Bromley, R.D.F. and Thomas, C.J. (eds) (1993a) *Retail Change: Contemporary Issues*. London, UCL Press.

Bromley, R.D.F. and Thomas, C.J. (1993b) 'The retail revolution, the carless shopper and disadvantage', *Transactions, Institute of British Geographers* 18: 222–36.

Burgess, J., Harrison, C.M. and Limb, M. (1988a) 'Exploring environmental values through the medium of small groups, Part 1: theory and practice', *Environment and Planning A* 20: 309–26.

Burgess, J., Harrison, C.M. and Limb, M. 1998b 'Exploring environmental values through the medium of small groups: Part 2: illustrations of a group at work', *Environment and Planning A* 20: 457–76.

Burrows, R. and Marsh, C. (eds) (1992) *Consumption and Class*, London: Macmillan.

Butler, J. (1990) *Gender Trouble: Feminism and the Subversion of Identity*, London: Routledge.

Callon, M. and Law, J. (1995) 'Agency and the hybrid collectif', *South Atlantic Quarterly* 94: 481–507.

Campbell, C. (1987) *The Romantic Ethic and the Spirit of Modern Consumerism*, Oxford: Blackwell.

Campbell, C. (1993) 'Shopping, pleasure and the context of desire', paper presented at the 4th International Conference on Consumption, Amsterdam.

Campbell, C. (1994) 'Understanding traditional and modern patterns of consumption in eighteenth century England: a character-action approach', in J. Brewer and R. Porter (eds) *Consumption and the World of Goods*, London: Routledge.

Campbell, C. (1996) 'The meaning of subjects and the meaning of actions: a cultural note on the sociology of consumption and theories of clothing', *Journal of Material Culture* 1: 93–106.

Carrier, J. (1995) *Gifts and Commodities: Exchange and Western Capitalism since 1770*, London: Routledge.

Carter, E., Donald, J. and Squires, J. (eds) (1993) *Space and Place: Theories of Identity and Location*, London: Lawrence & Wishart.

Catalano, A. (1996) 'Hey big spender . . .', *Estates Gazette* 960: 50–1.

Chaney, D. (1990) 'Dystopia in Gateshead: the Metrocentre as cultural form', *Theory, Culture and Society* 7: 49–68.

Chaney, D. (1996) *Lifestyles*, London: Routledge.

Chevalier, S. (1997) 'From woollen carpet to grass carpet: bridging house and garden in an English suburb', in D. Miller (ed) *Material Cultures*, London: UCL Press.

Clarke, A. (1997) 'Window shopping at home: classifieds, catalogues and new consumer skills', in D. Miller (ed) *Material Cultures*, London: UCL Press.

Collins, H.M. (1990) *Artificial Experts. Social Knowledge and Intelligent Machines*, Cambridge, MA: MIT Press.

Cook, I and Crang, P. (1996) 'The world on a plate: culinary culture, displacement and geographical knowledges', *Journal of Material Culture* 1: 131–53.

Cooke, L. and Wollen, P. (eds) (1995) *Visual Display. Culture Beyond Appearances*, Seattle: Bay Press.

Crang, P. (1995) 'It's showtime: on the workplace geographies of display in a restaurant in South-East England', *Environment and Planning D. Society and Space* 38: 219–46.

Crewe, L. and Gregson, N. (1997) 'The bargain, the knowledge and the spectacle: making sense of consumption in the space of the car boot sale', *Environment and Planning D. Society and Space* 15: 87–112.

Crewe, L., Gregson, N. and Longstaff, B. (1997) 'Excluded spaces of regulation: car boot sales as an enterprise culture out of control?', *Environment and Planning A* 29: 1717–37.

Crewe, L. and Lowe, M. (1996) 'Gap on the map? Towards a geography of consumption and identity', *Environment and Planning A* 27: 1877–98.

Crompton, R. (1993) *Class and Stratification*, Cambridge: Polity Press.

Davis, M. (1990) *City of Quartz*, London: Verso.

Debord, G. (1966) *The Society of the Spectacle*, Detroit: Black and Red.

de Certeau, M. (1984) *The Practice of Everyday Life*, Berkeley: University of California Press.

de Grazia, V. (ed.) (1996) *Sex From Things*, Berkeley: University of California Press.

Denzin, N.K. (1993) *The Research Act: A Theoretical Introduction to Sociological Methods*, Chicago: Aldine.

DeVault, M. (1991) *Feeding the Family*, Chicago: University of Chicago Press.

Department of the Environment (1993) *Town Centres and Retail Developments*, Planning Policy Guidance 6 (revised), London: Department of the Environment.

Department of the Environment (1994) *Transport*, Planning Policy Guidance 13, London: Department of the Environment.

Douglas, M. (1969) *Purity and Danger*, London: Routledge and Kegan Paul.

Douglas, M. (1996) *Thought Styles: Critical Essays on Good Taste*, London: Sage.

Douglas, M. and Isherwood, B. (1979) *The World of Goods*, London: Allen Cone.

du Gay, P. (1996) *Consumption and Identity at Work*, London: Sage.

Evernden, B. (1992) *The Social Creation of Nature*, Baltimore: Johns Hopkins Press.

Featherstone, M. (1990) *Consumer Culture and Postmodernism*, London: Sage.

Feinberg, R.A., Scheffler, B., Meoli, J. and Rummel, A. (1989) 'There's something social happening at the mall', *Journal of Business Psychology* 4: 49–63.

Finch, J. (1986) *Family Obligations and Social Change*, Cambridge: Polity Press.

Finch, J. and Mason, J. (1990) 'Divorce, remarriage and family obligations', *Sociological Review* 38: 219–46.

Fischer, E. and Arnold, S.J. (1990) 'More than a labour of love: gender roles and Christmas gift shopping', *Journal of Consumer Research* 17: 333–45.

Fiske, J. (1989) *Reading the Popular*, London: Unwin Hyman.

Flanders, A., Pomeranz, R. and Woodward, J. (1968) *Experiments in Industrial Democracy*, London: Faber & Faber.

Frankenberg, R. (1993) *White Women, Race Matters: The Social Construction of Whiteness*, London: Routledge.

Fraser, W. (1981) *The Coming of the Mass Market, 1850–1914*, London: Hamish Hamilton.

Friedberg, A. (1993) *Window Shopping: Cinema and the Postmodern*, Berkeley: University of California Press.

Fyfe, N.R. and Bannister, J. (1995) 'The eyes of the street: surveillance, citizenship and the city', paper presented at the annual conference of the Institute of British Geographers, University of Northumbria at Newcastle (January).

Game, A. (1991) *Undoing the Social. Towards a Deconstructive Sociology*, Milton Keynes: Open University Press.

Gardner, C. and Sheppard, J. (1989) *Consuming Passions: The Rise of Retail Culture*, London: Unwin Hyman.

Gates, H.L. (1986) *'Race', Writing and Difference*, Chicago: University of Chicago Press.

Ghosh, A. (1992) *In an Antique Land*, London: Granta Books.

Giddens, A. (1991) *Modernity and Self-Identity: Self and Society in the Late-Modern Age*, Cambridge: Polity Press.

Gillespie, M. (1995) *Television, Ethnicity and Cultural Change*, London: Routledge.

Gilroy, P. (1987) *There Ain't No Black in the Union Jack*, London: Hutchinson.

Gilroy, P. (1993a) *The Black Atlantic*, London: Verso.

Gilroy, P. (1993b) 'It ain't where you're from, it's where you're at', in *Small Acts*, London: Serpent's Tail, 120–45.

Glennie, P. (1995) 'Consumption within historical studies', in D. Miller (ed.) *Acknowledging Consumption. A Review of New Studies*, London: Routledge, 164–203.

Glennie, P. and Thrift, N.J. (1992) 'Modernity, urbanism and modern consumption', *Environment and Planning D. Society and Space* 10: 423–43.

Glennie, P. and Thrift, N.J. (1996) 'Consumers, identities, and consumption spaces in early modern England', *Environment and Planning A* 28: 25–46.

Goldman, A.E., McDonald, S. and Schwartz, S. (1987) 'A theoretical foundation for group interview techniques', in *The Group Depth Interview: Principles and Practice*, Englewood Cliffs: Prentice Hall.

Goss, J. (1993) 'The "magic of the mall": an analysis of form, function, and meaning in the contemporary retail built environment', *Annals of the Association of American Geographers* 83: 18–47.

Grosz, E. (1994) *Volatile Bodies*, Bloomington: Indiana University Press.

Hall, C. (1992) *White, Male and Middle Class: Essays in Feminism and History*, Cambridge: Cambridge University Press.

Hall, S. (1992a) 'New ethnicities', in J. Donald and A. Rattansi (eds) *'Race', Culture and Difference*, London: Sage, 252–9.

Hall, S. (1992b) 'The question of cultural identity', in S. Hall, D. Held and T. McGrew (eds) *Modernity and its Futures*, Cambridge: Polity Press, 273–316.

Haringey in Figures (November 1993) London: London Borough of Haringey.

Haynes, A. (1983) 'Murderous millinery: the struggle for the Plumage Act, 1921', *History Today* 33 (July).

Hebdige, D. (1979) *Subculture: The Meaning of Style*, London: Methuen.

Hermes, J. (1993) 'Media, meaning and everyday life', *Cultural Studies* 7: 493–506.

Hobsbawm, E. (1996) 'Identity politics and the Left', *New Left Review* 217: 38–47.

Holbrook, B. and Jackson, P. (1996a) 'Shopping around: focus group research in North London', *Area* 28: 136–42.

Holbrook, B. and Jackson, P. (1996b) 'The social milieux of two North London shopping centres', *Geoforum* 27: 193–204.

Hopkins, J.S.P. (1991) 'West Edmonton Mall: landscape of myths and elsewhereness', *Canadian Geographer* 34: 2–17.

Hopkins, J.S.P. (1994) 'A consumption of resistance: challenging the socio-legal identities of malls', paper presented to the Association of American Geographers annual meeting, San Francisco.

House of Commons Select Committee on the Environment (1994) *Shopping Centres and their Future*, Fourth Report (1993–4 Session), London: HMSO (October).

Howard, E.B. (1989) *Prospects for Out-of-Town Retailing: The Metro Experience*, Harlow: Longman.

Howard, E.B. and Davies, R.L. (1992) *Meadowhall: The Impact of One Year's Development*, OXIRM Research Paper D9, Oxford: Oxford Institute of Retail Management.

201

Howes, D. (ed.) (1996) *Cross-Cultural Consumption: Global Markets, Local Realities,* London: Routledge.

Jackson, P. (1993) 'Towards a cultural politics of consumption', in J. Bird, B. Curtis, T. Putnam, G. Robinson and L. Tickner (eds) *Mapping the Futures: Local Cultures, Global Change,* London: Routledge, 207–28.

Jackson, P. (1994) 'Black male: advertising and the cultural politics of masculinity', *Gender, Place and Culture* 1: 49–59.

Jackson, P. (1998) 'Domesticating the street: the contested spaces of the high street and the mall', in N. Fyfe (ed.) *Images of the Street,* London: Routledge.

Jackson, P. and Holbrook, B. (1995) 'Multiple meanings: shopping and the cultural politics of identity', *Environment and Planning A* 27: 1913–30.

Jackson, P. and Thrift, N. (1995) 'Geographies of consumption', in D. Miller (ed.) *Acknowledging Consumption,* London: Routledge, 204–37.

Jencks, C. (1991) *The Language of Post-Modern Architecture,* London: London Academy.

Jacobs, J. (1961) *The Death and Life of Great American Cities: The Failure of Town Planning,* New York: Random House (reprinted 1965 Harmondsworth: Penguin).

Journal of Material Culture (1996) 'Editorial', *Journal of Material Culture* 1: 5–14.

Kearns, G. and Philo, C. (eds) (1993) *Selling Places: The City as Cultural Capital, Past and Present,* Oxford: Pergamon, 211–30.

Keith, M. (1993) *Race, Riots and Policing,* London: UCL Press.

Keith, M. and Pile, S. (eds) (1993) *Place and the Politics of Identity,* London: Routledge.

Kelly, G. (1955) *The Psychology of Personal Constructs,* New York: W.W. Norton.

Kitzinger, J. (1994) 'The methodology of focus groups: the importance of interaction between research participants', *Sociology of Health and Illness* 16: 103–26.

Kowinski, W.S. (1985) *The Malling of America: An Inside Look at the Great Consumer Paradise,* New York: William Morrow & Co.

Lamont, M. (1992) *Money, Morals and Manners,* Chicago: University of Chicago Press.

Lancaster, B. (1995) *The Department Store: A Social History,* Leicester: Leicester University Press.

Lash, S. and Friedman, J. (eds) (1992) *Modernity and Identity,* Oxford: Basil Blackwell.

Latour, B. (1992) 'Where are the missing masses? A sociology of a few mundane artefacts', in W.E. Bijker and J. Law (eds) *Shaping Technology/Building Society,* Cambridge, MA: MIT Press, 235–58.

Lave, J. (1986) 'The values of quantification', in J. Law (ed.) *Power, Action and Belief. A New Sociology of Knowledge,* London: Routledge and Kegan Paul, 234–63.

Lave, J. and Wenger, E. (1991) *Situated Learning. Legitimate Peripheral Participation,* Cambridge: Cambridge University Press.

Lawrence, E. (1992) 'Just plain common sense: the 'roots' of racism', in Centre for Contemporary Cultural Studies (ed.) *The Empire Strikes Back: Race and Racism in 70s Britain,* London: Hutchinson, 47–94.

Lévi-Strauss, C. (1972) *The Savage Mind,* London: Weidenfeld & Nicolson.

Lord, J.D. (1985) 'The malling of the American landscape', in J.A. Dawson and J.D. Lord (eds) *Shopping Centre Development: Policies and Prospects,* London: Croom Helm.

Lowe. M. (1993) 'Local hero! an examination of the role of the regional entrepreneur in the regeneration of Britain's regions', in G. Kearns and C. Philo (eds) *Selling Places,* Oxford: Pergamon, 211–30.

Lowe, M. and Crewe, L. (1996) 'Shopwork: image, customer care and the restructuring of retail employment', in N. Wrigley and M. Lowe (eds) *Retailing, Consumption and Capital: Towards the New Retail Geography,* London: Longman, 196–207.

Lowe, M. and Wrigley, N. (1996) 'Towards the new retail geography', in N. Wrigley and M. Lowe (eds) *Retailing, Consumption and Capital. Towards the New Retail Geography,* London: Longman, 3–30.

Lunt, P. and Livingstone, S. (1992) *Mass Consumption and Personal Identity*, Buckingham: Open University Press.

Macfarlane, A. (1987) *The Culture of Capitalism*, Oxford: Basil Blackwell.

Maffesoli, M. (1996) *The Times of the Tribes* (trans. D. Smith), London: Sage.

Marsden, T. and Wrigley, N. (1995) 'Regulation, retailing and consumption', *Environment and Planning A* 27: 1899–1912.

Marsden, T. and Wrigley N. (1996) 'Retailing, the food system and the regulatory state', in N. Wrigley and M. Lowe (eds) *Retailing, Consumption and Capital*, Harlow: Longman.

Massey, D. (1994) *Space, Place and Gender*, Cambridge: Polity Press.

McClintock, A. (1995) *Imperial Leather: Race, Gender and Sexuality in the Colonial Contest*, London: Routledge.

McGoldrick, P. and Thompson, M. (1992) *Regional Shopping Centres: Out-of-town versus In-town*, Aldershot: Avebury Press.

McKendrick, N., Brewer, J. and Plumb, J. (1982) *The Birth of a Consumer Society. The Commercialisation of Eighteenth Century England*, London: Hutchinson.

McRobbie, A. (1994) 'Folk devils fight back', *New Left Review* 203: 107–16.

Miles, M.B. and Huberman, A.M. (1984) *Qualitative Data Analysis: A Sourcebook of New Methods*, London: Sage.

Miller, D. (1987) *Mass Consumption and Material Culture*, Oxford: Blackwell.

Miller, D. (1988a) 'Modernism and suburbia as contemporary ideology', in D. Miller and C. Tilley (eds) *Ideology, Power and Prehistory*, Cambridge: Cambridge University Press.

Miller, D. (1988b) 'Appropriating the state on the council estate', *Man* 23: 353–72.

Miller, D. (1992) 'Personal communication', in R. Silverstone and E. Hirsch (eds) *Consuming Technologies*, London: Routledge, 225.

Miller, D. (ed.) (1993) *Unwrapping Christmas*, Oxford: Oxford University Press.

Miller, D. (1994) *Modernity: An Ethnographic Approach*, Oxford: Berg.

Miller, D. (ed.) (1995) *Acknowledging Consumption: A Review of New Studies*, London: Routledge.

Miller, D. (1997) *Capitalism: An Ethnographic Approach*, Oxford: Berg.

Miller, D. (forthcoming) *A Theory of Shopping*, Cambridge: Polity Press.

MINTEL (1994) *Retail Intelligence. Consumer Shopping Habits*, London: MINTEL.

Mitchell, J.C. (1983) 'Case and situation analysis', *Sociological Review* 31: 187–211.

Mitchell, W.T.J. (1995) *City of Bits. Space, Place and the Infobahn*, Cambridge, MA: MIT Press.

Modood, T. (1988) '"Black", racial equality and Asian identity', *New Community* 14: 397–404.

Morris, M. (1988) 'Things to do with shopping centres', in S. Sheridan (ed.) *Grafts. Feminist Cultural Criticism*, London: Verso, 193–225.

Mort, F. (1989) 'The politics of consumption', in S. Hall and M. Jacques (eds) *New Times: The Changing Face of Politics in the 1990s*, London: Lawrence & Wishart, 160–72.

Mort, F. (1996) *Cultures of Consumption: Masculinities and Social Space in Late Twentieth Century Britain*, London: Routledge.

Nava, M. (1992) *Changing Cultures: Feminism, Youth and Consumerism*, London: Routledge.

Nava, M. (1996) 'Modernity's disavowal: women, the city and the department store', in M. Nava and A. O'Shea (eds) *Modern Times. Reflections on a Century of English Modernity*, London: Routledge, 38–76.

Nead, L. (1997) 'Mapping the self: gender, space and modernity in mid-Victorian London', *Environment and Planning A* 29: 659–72.

O'Brien, L. and Harris, F. (1991) *Retailing: Shopping, Society, Space*, London: David Fulton.

Pahl, R. (1995) *After Success. Fin de Siècle Anxiety and Identity*, Cambridge: Polity Press.

Parry, J. and Bloch, M. (1989) *Money and the Morality of Exchange*, Cambridge: Cambridge University Press.

Phillips, A. (1994) *On Flirtation*, London: Faber & Faber.

Philips, H.C. and Bradshaw, R.P. (1993) 'How customers actually shop: customer interaction with the point of sale', *Journal of the Market Research Society* 35: 51–62.

Pile, S. and Thrift, N.J. (eds) (1995) *Mapping the Subject. Geographies of Cultural Transformation*, London: Routledge.

Pred, A. (1995) *Recognizing European Modernities: A Montage of the Present*, London: Routledge.

Price, J. (1995) 'Looking for nature at the mall: a field guide to the Nature Company', in W. Cronon (ed.) *Uncommon Ground*, Cambridge, MA: Harvard University Press, 186–203.

Pringle, R. (1983) 'Women and consumer capitalism', in C.V. Baldock and B. Cass (eds) *Women, Social Welfare and the State in Australia*, Sydney: Allen and Unwin.

Probyn, E. (1996) *Outside Belongings*, London: Routledge.

Prus, R. and Dawson, L. (1991) '"Shop Til You Drop": shopping as recreational and laborious activity', *Canadian Journal of Sociology* 16: 145–64.

Raven, H., Lang, T. and Dumonteil, C. (1995) *Off Our Trolleys? Food Retailing and the Hypermarket Economy*, London: Institute for Public Policy Research.

Reekie, G. (1993) *Temptations. Sex, Selling and the Department Store*, Sydney: Allen and Unwin.

Regional Trends (1994) London: HMSO.

Relph, E.C. (1976) *Place and Placelessness*, London: Pion.

Retail Planning Associates (1978) *The Brent Cross Regional Shopping Centre: Characteristics and Early Effects*, Uxbridge: Middlesex Polytechnic.

Reynolds, J. and Howard, E. (1993) *The UK Regional Shopping Centre: The Challenge for Public Policy*, OXIRM Research Paper A28, Oxford: Oxford Institute for Retail Management.

Robillard, A. (1996) 'Anger-in-the-social order', *Body and Society* 2: 17–30.

Rowlands, M. (1995) 'Inconsistent temporalities in a nation-space', in D. Miller (ed.) *Worlds Apart*, London: Routledge, 23–42.

Rowley, G. (1993) 'Prospects for the Central Business District', in R.D.F. Bromley and C.J. Thomas (eds) *Retail Change: Contemporary Issues*, London: UCL Press.

Rutherford, J. (ed.) (1990) *Identity: Community, Culture, Difference*, London: Lawrence & Wishart.

Said, E. (1993) *Culture and Imperialism*, London: Chatto & Windus.

Samuel, R. (1994) *Theatres of Memory*, London: Verso.

Savage, M., Barlow, J., Dickens, P. and Fielding, T. (1992) *Property, Bureaucracy and Culture*, London: Routledge.

Savage, M. and Warde, A. (1993) *Urban Sociology, Capitalism and Modernity*, London: Macmillan.

Schama, S. (1987) *The Embarrassment of Riches*, London: Fontana.

Sennett, R. (1977) *The Fall of Public Man*, London: Faber & Faber.

Sennett, R. (1990) *The Conscience of the Eye: The Design and Social Life of Cities*, London: Faber & Faber.

Shields, R. (ed.) (1992a) *Lifestyle Shopping: The Subject of Consumption*, London: Routledge.

Shields, R. (1992b) 'The individual, consumption cultures and the fate of community', in R. Shields (ed.) *Lifestyle Shopping: The Subject of Consumption*, London: Routledge, 99–113.

Slater, D. (1993) 'Going shopping: markets, crowds and consumption', in C. Jencks (ed.) *Cultural Reproduction*, London: Routledge.

Slater, D. (1997) *Consumer Culture and Modernity*, Cambridge: Polity Press.

Smith, N. (1996) 'The production of nature', in G. Robertson, M. Mash, L. Tickner, J. Bird, B. Curtis and T. Putnam (eds) *Futurenatural: Nature, Science, Culture*, London: Routledge, 35–54.

Smith, A. (1986) *The Wealth of Nations*, London: Penguin Classics.

Smith, S.J. (1985) 'News and the dissemination of fear', in J. Burgess and J.R. Gold (eds) *Geography, the Media and Popular Culture*, London: Croom Helm, 229–53.

Smith, S.J. (1987) 'Fear of crime: beyond a geography of deviance', *Progress in Human Geography* 11: 1–23.

Somers, M.R. (1994) 'The narrative constitution of identity: a relational and network approach', *Theory and Society* 23: 605–49.

Stansell, C. (1986) *City of Women: Sex and Class in New York, 1789–1860*, New York: Alfred Knopf.

Stewart, D.W. and Shamdasani, P.N. (1992) *Focus Groups: Theory and Practice*, London: Sage.

Strathern, M. (1996) 'Enabling identity', in S. Hall and P. du Gay (eds) *Questions of Cultural Identity*, London: Sage, 37–52.

Swanson, G. (1995) '"Drunk with the Glitter": consuming spaces and sexual geographies', in S. Watson and K. Gibson (eds) *Postmodern Cities and Spaces*, Oxford: Basil Blackwell, 80–98.

Sykes, W. (1991) 'Taking stock: issues from the literature on validity and reliability in qualitative research', *Journal of the Market Research Society* 33: 3–12.

Taub, R.P., Taylor, D.G. and Dunham, J.D. (1984) *Paths of Neighborhood Change*, Chicago: University of Chicago Press.

Taussig, M. (1992) *The Nervous System*, London: Routledge.

Taylor, I., Evans, K. and Fraser, P. (1996) *A Tale of Two Cities: A Study in Manchester and Sheffield*, London: Routledge.

Taylor, P. (1991) 'The English and their Englishness: "a curiously mysterious, elusive and little understood people"', *Scottish Geographical Magazine* 107: 146–61.

Thomas, C. and Bromley, R.D.F. (1993) 'The impact of out-of-centre retailing, in R.D.F. Bromley and C. Thomas (eds) *Retail Change: Contemporary Issues*, London: UCL Press, 126–152.

Thomas, C. and Bromley, R.D.F. (1996) 'Safety and shopping: peripherality and shopper anxiety in the city centre', *Environment and Planning C: Government and Policy* 14: 469–88

Thomas, K. (1983) *Man and the Natural World: Changing Attitudes to Nature in England 1500– 1800*, London: Allen Lane.

Thrift, N.J. (1996) *Spatial Formations*, London: Sage.

UDAG (Urban Design Action Group) (1990) untitled, London: Wood Green.

URBED (Urban and Economic Development Group) (1994) *Vital and Viable Town Centres: Meeting the Challenge*, Report to the Department of the Environment by the Urban and Economic Development Group, London: HMSO.

Valentine, G. (1989) 'The geography of women's fear', *Area* 21: 385–90.

Venkatesh, A. (1995) 'Ethnoconsumerism: a new paradigm to study cultural and cross-cultural consumer behaviour', in J. Costa and G. Bamossy (eds) *Marketing in a Multicultural World*, London: Sage, 26–67.

Walkowitz, J. (1992) *City of Dreadful Delight: Narratives of Sexual Danger in Late-Victorian London*, London: Virago.

Walzer, M. (1986) 'Pleasures and costs of urbanity', *Dissent*: 470–76.

Warde, A. (1994) 'Consumption, identity-formation and uncertainty', *Sociology* 23: 877–98.

Westlake, T. (1993) 'The disadvantaged consumer: problems and policies', in R.D.F. Bromley and C. Thomas (eds) *Retail Change: Contemporary Issues*, London: UCL Press, 172–91.

Weston, K. (1991) *Families We Choose: Lesbians, Gays, Kinship*, New York: Columbia University Press.

Whitehead, A. (1993) 'Regional shopping centres: a rejoinder', *Local Government Policy Making* 20: 50–9.

Williams, C.C. (1995) 'Opposition to regional shopping centres in Great Britain: a clash of cultures?' *Journal of Retailing and Consumer Services* 2: 241–9.

Williams, R. (1973) *The Country and the City*, London: Chatto & Windus.

Williams, R. (1976) *Keywords*, London: Fontana.

Williams, R.H. (1982) *Dream Worlds: Mass Consumption in Late Nineteenth-Century France*, Berkeley and Los Angeles: University of California Press.

Williamson, J. (1986) *Consuming Passions*, London: Marion Boyars.

Williamson, J. (1992) 'I-less and gaga in the West Edmonton Mall: towards a pedestrian feminist reading', in D.H. Currie and V. Raoul (eds) *The Anatomy of Gender: Women's Struggles for the Body*, Ottawa: Carleton University Press, 97–115.

Willis, P. (1975) *Profane Culture*, London: Macmillan.

Willis, P. (1990) *Common Culture: Symbolic Work at Play in the Everyday Cultures of the Young*, Milton Keynes: Open University Press.

Willis, S. (1991) *A Primer of Daily Life*, London: Routledge.

Wilson, E. (1992) 'The invisible flâneur', *New Left Review* 191: 90–110.

Woodward, J. (1960) *The Saleswoman. A Study of Attitudes and Behaviour in Retail Distribution*, London: Pitman.

Woods, P. (1991) *In Times Past: Wood Green and Tottenham with West Green and Haringey*, London: Hornsey Historical Society.

Wrigley, N. (1991) 'Is the golden age of British grocery retailing at a watershed?' *Environment and Planning A* 23: 1537–44.

Wrigley, N. (1992) 'Antitrust regulation and the restructuring of grocery retailing in Britain and the USA', *Environment and Planning A*, 24: 727–49.

Wrigley, N. (1997a) 'British food retail capital in the USA, 1: Sainsbury and the Shaw's experience', *International Journal of Retail and Distribution Management* 25.

Wrigley, N. (1997b) 'British food retail capital in the USA, 2: Giant prospects?' *International Journal of Retail and Distribution Management* 25.

Wrigley, N. and Lowe, M. (eds) (1996) *Retailing, Consumption and Capital. The New Retail Geography*, Harlow: Longman.

Young, R.J.C. (1995) *Colonial Desire: Hybridity in Theory, Culture and Race*, London: Routledge.

Zukin, S. (1995) *The Cultures of Cities*, Oxford: Basil Blackwell.

INDEX